Book T1

PUBLISHED BY THE PRESS SYNDICATE OF THE UNIVERSITY OF CAMBRIDGE
The Pitt Building, Trumpington Street, Cambridge, United Kingdom

CAMBRIDGE UNIVERSITY PRESS
The Edinburgh Building, Cambridge CB2 2RU, UK
40 West 20th Street, New York, NY 10011–4211, USA
10 Stamford Road, Oakleigh, Melbourne 3166, Australia
Ruiz de Alarcón 13, 28014 Madrid, Spain
Dock House, The Waterfront, Cape Town 8001, South Africa

http://www.cambridge.org

© The School Mathematics Project 2001
First published 2001

Printed in the United Kingdom at the University Press, Cambridge

Typeface Minion   *System* QuarkXPress®

A catalogue record for this book is available from the British Library

ISBN 0 521 79858 2 paperback

Typesetting and technical illustrations by The School Mathematics Project
Other illustrations by Robert Calow and Steve Lach at Eikon Illustration
Cover image © Tony Stone Images/Darryl Torckler
Cover design by Angela Ashton

The publishers would like to thank the following for supplying photographs:
page 41   GreatBuildings.com photos © Artifice, Inc. (*left*, Habitat '67; *right*, Guggenheim Museum, Bilbao)
page 45   Egon Publishers
page 54   © Charles O'Rear/CORBIS
page 59   © National Gallery, London (Hans Holbein the Younger, detail from *Jean de Dinteville and Georges de Selve* 'The Ambassadors')
page 85   National Meteorological Library © J. F. P. Galvin (0 and 4 oktas); © C. S. Broomfield (8 oktas) Paul Scruton (7 oktas)
page 89   Colin Benwell
page 121   National Aeronautics and Space Administration (USA)
page 122   National Aeronautics and Space Administration (USA)
All other photographs by Graham Portlock

Logos on page 15 appear courtesy of *A*, General Motors Corporation; *B*, BMW (GB) Ltd; *D*, Honda Motor Corporation; *E*, Mercedes-Benz, DaimlerChrysler; *G*, Mitsubishi Motors; *H*, Suzuki Motor Corporation; *I*, Toyota Motor Corporation.

Catalogue spread on page 64 appears courtesy of Littlewood's.

Platform ticket on page 82 is ©The Edinburgh Woollen Mill Limited.

The authors and publishers would like to thank the staff and pupils of Impington Village College, Cambridge, for their help with the production of this book.

NOTICE TO TEACHERS
It is illegal to reproduce any part of this work in material form (including photocopying
and electronic storage) except under the following circumstances:
(i) where you are abiding by a licence granted to your school by the Copyright Licensing Agency;
(ii) where no such licence exists, or where you wish to exceed the terms of a licence, and you
have gained the written permission of Cambridge University Press;
(iii) where you are allowed to reproduce without permission under the provisions of Chapter 3
of the Copyright, Designs and Patents Act 1988.

Contents

1 Time — 4
2 Action and result puzzles — 8
3 Chance — 9
4 Symmetry — 15
5 Decimals 1 — 20
   Review 1 — 28
6 Number grids — 30
7 Stars and angles — 40
8 Lines at right angles — 41
9 Comparisons — 45
10 Ice cream — 53
11 Parallel lines — 54
12 Anamorphs — 59
   Review 2 — 60
13 Practical problems — 62
14 Angle dominoes — 65
15 Quadrilaterals — 66
16 Is it an add? — 72
17 Desk tidy — 78
18 Frequency — 79
19 Amazing but true! — 89
   Review 3 — 90
20 Photo display — 92
21 Fractions 1 — 93
22 Enlargement — 98
23 Calculate in order — 104
24 Graphs and charts — 110
25 Fractions 2 — 117
26 Negative numbers — 121
   Review 4 — 127

1 Time

This work will help you calculate with time.

A Happiness graphs

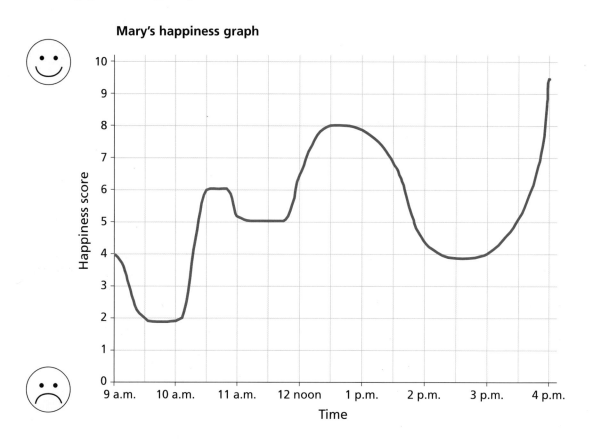

B Time planner

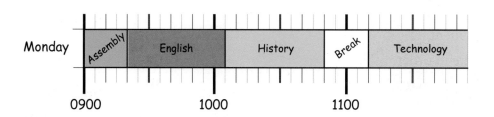

C At the same time

D Time lines

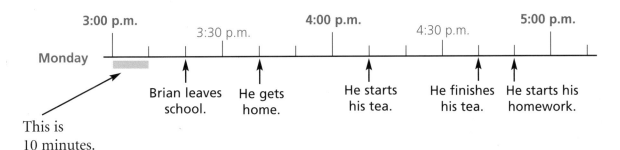

D1 What time does Brian leave school?

D2 What time does he get home?

D3 How long does it take him to get home?

D4 What time does he start his tea?

D5 How long does he take to eat his tea?

D6 What time does he start his homework?

D7 Brian spends 40 minutes on his homework. What time does he finish it?

D8 How long is it between
(a) getting home and starting his tea
(b) getting home and starting his homework

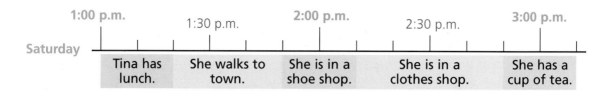

D9 Tina starts lunch at 1 o'clock. What time does she finish lunch?

D10 How long does it take Tina to walk to town?

D11 How long does she spend altogether in the shoe shop and the clothes shop?

D12 How long is it between finishing her lunch and having a cup of tea?

D13 After her cup of tea, Tina walks home. She takes 35 minutes to get home.
What time does she get home?

E How long?

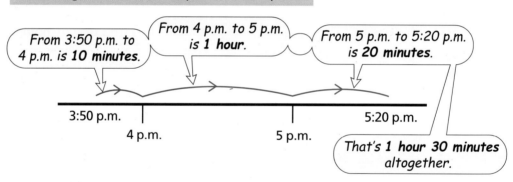

E1 How long is it
(a) from 4:30 p.m. to 6:10 p.m.
(b) from 2:40 p.m. to 5:10 p.m.
(c) from 8:20 a.m. to 11:10 a.m.
(d) from 11:30 a.m. to 1:20 p.m.

E2 How long is it
(a) from a quarter to 7 to half past 8
(b) from half past 2 to 10 past 4
(c) from 25 past 9 to 10 past 11
(d) from 10 past 8 to 20 to 10

E3 A film starts at 9:30 p.m. and finishes at 11:10 p.m. How long does it last?

E4 Sasha wants to go to a film. It starts at 8:20 p.m.
She leaves home at a quarter to 8.
How long has she got before the film starts?

Buses leave here at
8:15 a.m.
9:45 a.m.
11:10 a.m.
1:15 p.m.
4:05 p.m.

E5 Gary gets to the bus stop at 10:55 a.m.
How long does he have to wait for a bus?

E6 Sue gets to the bus stop at 12:40 p.m.
How long does she have to wait for a bus?

E7 Misha gets to the bus stop at a quarter to 4.
How long does he have to wait for a bus?

E8 Paul wants to see *Home Alone* at 2:15 p.m.
It takes him 30 minutes from home to the cinema.

What is the latest time he can leave home?

E9 Lana wants to see *Home Alone* at 4:10 p.m.
It will take her 45 minutes to get there.

What is the latest time she can leave?

MASCOT Cinema

Home Alone

| 2:15 p.m. | 4:10 p.m. |
| 6:20 p.m. | 8:45 p.m. |

E10 Colin leaves home at 7:45 p.m.
It takes him 25 minutes to get to the cinema.

When he gets there, how long will he have to wait for the film to start?

E11 (a) How long does the bus take to get from Market Hill to Red Lion?

(b) How long does it take to get from West Gate to Flybridge?

E12 Roddy is at West Gate at half past 8.
How long does he have to wait for a bus?

E13 The bus is a quarter of an hour late when it gets to Flybridge.

What time does it arrive at Flybridge?

TURBO BUS timetable	
Market Hill	8:10
Castle St	8:25
West Gate	8:40
York Road	9:00
Red Lion	9:25
Tay Cross	9:45
Oak Hill	10:05
Flybridge	10:20

What progress have you made?

Statement

I can work out how long it is between two times.

Evidence

1 How long is it
   (a) from half past 6 to a quarter past 7
   (b) from 2:15 p.m. to 3:35 p.m.
   (c) from 8:50 a.m. to 10:10 a.m.
   (d) from 25 past 2 to 10 past 4

Action and result puzzles

These puzzles involve adding, subtracting, multiplying and dividing.
Doing them will help you

- carry out these sorts of calculations in your head
- understand more about what happens when you do these calculations
- explain your methods to other people and listen to their explanations

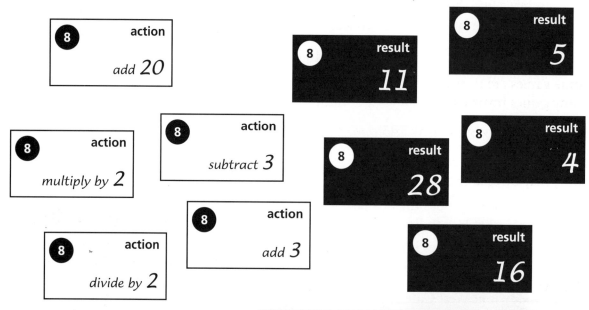

- How do you match the cards?

There are several different puzzles.
Try to solve some of them.

Try making up your own puzzles.

3 Chance

This is about games and other situations where the outcome is uncertain because it is a matter of chance.
The work will help you
- decide whether a game is fair or unfair
- understand the probability scale
- write a probability as a fraction

A Chance or skill?

Some games are games of skill. Some are games of chance. Many games involve both chance and skill.

There are three games on sheets 111, 112 and 113.

Play each game and decide if they are games of skill or chance. Some may involve both skill and chance.

Fours

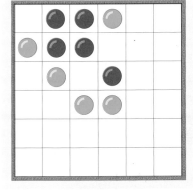

Line of three

1	3	●	4	3	3
2	●	3	2	2	2
4	2	1	1	●	6
1	●	6	6	5	3
5	5	5	4	●	5
4	4	●	1	4	6

Jumping the line

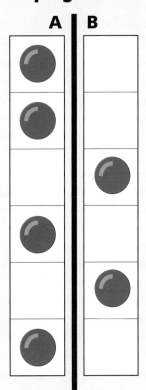

B Fair or unfair?

Three way race

For three players (A, B and C)

Each puts a counter at the start of the track.

Two dice are rolled.

If both numbers are even, A moves forward one space.

If both numbers are odd, B moves forward one space.

If one number is even and one odd, C moves forward one space.

The first to get to the end of the track is the winner.

	Finish	
	Start	
A	B	C

- Play the game several times.
  Keep a record of who wins (A, B or C).

Winner	Tally	Number of wins
A	ĦĦ I	
B	III	
C		

- Is it a fair game? Do A, B and C all have the same chance of winning?

Rat races

Teacher-led games

Play the two rat race games with your teacher.

Are these games fair?
Does every rat have an equal chance of winning?

C Probability

Probability is a way of saying how likely something is.

Something which never happens has probability 0.
Something which is certain to happen has probability 1.

Things which have a chance of happening have probabilities between 0 and 1.

This is a probability washing line.

Where would you hang these on the line?

- The probability that a coin lands heads
- The probability that Rat 1 wins the second rat race
- The probability that the sun will rise tomorrow morning
- The probability that a particular ticket wins the National Lottery

Make up some more examples.

C1 Draw a probability scale like this.

```
0            1/2            1
|-------------|-------------|
```

Mark these roughly on your scale with arrows.

(a) The probability that when you roll a dice you get an even number

(b) The probability that when you roll a dice you get a 6

C2 Out of every 1000 babies born, 515 are boys and 485 are girls.

(a) Is a new-born baby more likely to be a boy or a girl?

(b) On the scale you drew, mark roughly the probability that a new-born baby will be a boy.

C3 Which event below goes with each of the arrows on the probability line?

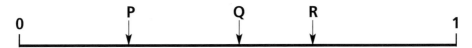

(a) Choosing a heart from a shuffled pack of cards

(b) Getting 1, 2, 3 or 4 when you roll a dice

(c) Choosing a red pen from a pencil case with 5 red and 5 black pens

D Spinners

Sometimes a spinner is used instead of a dice.

This spinner has five equal sections.
You spin the arrow. When it stops, it points to a colour.

The spinner is fair.
Every section has the same probability of winning.

Suppose you have chosen yellow.
The probability that yellow will win is $\frac{1}{5}$.

Two sections are coloured red.
The probability that red will win is $\frac{2}{5}$.

D1 Match the cards here with the probability that red will win on each of these spinners. $\frac{1}{8}$ $\frac{1}{6}$ $\frac{1}{4}$ $\frac{1}{2}$

(a) 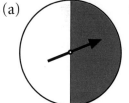 (b) 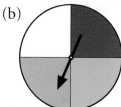 (c) 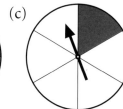 (d)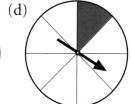

D2 Which of these fractions shows the probability that green will win on this spinner? $\frac{2}{3}$ $\frac{1}{5}$ $\frac{2}{5}$

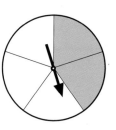

D3 What is the probability that blue will win on each of these spinners?

(a) 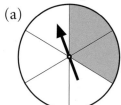 (b) 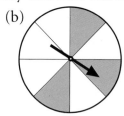 (c) 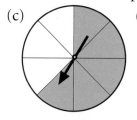 (d)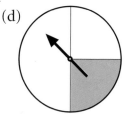

D4 With this spinner, what is the probability that
(a) yellow wins
(b) blue wins
(c) white wins

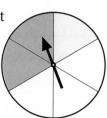

D5 With this spinner, what is the probability that
  (a) red wins
  (b) blue wins
  (c) white wins

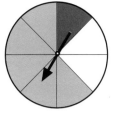

D6 With this spinner, what is the probability that red will **not** win?

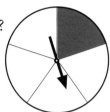

D7 What is the probability that red will **not** win on each of these spinners?
  (a) 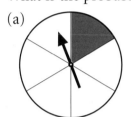 (b) 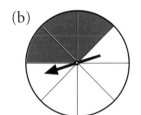 (c)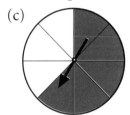

D8 What is the probability that red **will** win on each of these spinners?
  (a) 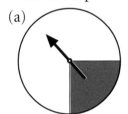 (b) 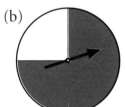 (c)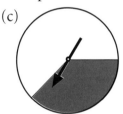

Make your own spinner

Draw a circle and divide it into equal sections. You can colour the sections or number them.

Put a pencil or compass point through a paper clip, so that the point is at the centre of the circle.

Spin the paper clip and see where it stops.
  • How can you check if your spinner is fair?

What progress have you made?

Statement

I can find out by experiment whether a game of chance is fair or unfair.

Evidence

1. Find out if this game is fair.

   There are two players, A and B.
   They take turns to throw two dice.
   If both numbers are less than 4, A wins a point.
   If not, then B wins a point.

   The first to get 10 points wins the game.

Statement

I understand the probability scale.

Evidence

2. Which letter on this probability scale shows the probability of each of the events below?

   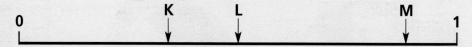

   (a) Choosing a red card (heart or diamond) from a pack of playing cards
   (b) Choosing a red sweet from a bag with 7 red and 1 yellow sweet
   (c) Rolling a 5 or a 6 with a dice

Statement

I can write a probability as a fraction.

Evidence

3. On this spinner, what is the probability that
   (a) blue will win
   (b) blue will not win
   (c) red will win
   (d) red will not win

4 Symmetry

This work will help you
- recognise rotation symmetry
- draw patterns with rotation symmetry
- find all the different symmetries in a shape

A What is symmetrical about these shapes?

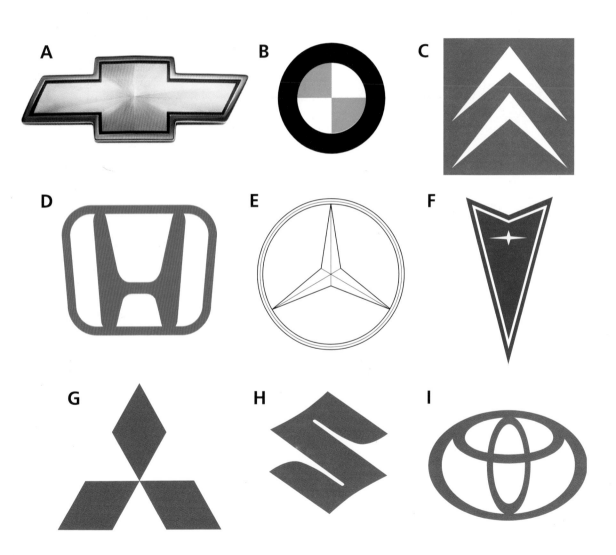

B Rotation symmetry

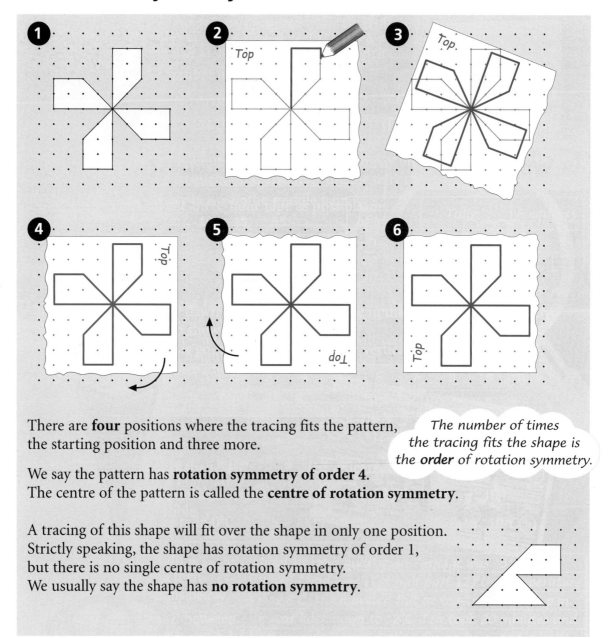

There are **four** positions where the tracing fits the pattern, the starting position and three more.

*The number of times the tracing fits the shape is the **order** of rotation symmetry.*

We say the pattern has **rotation symmetry of order 4**.
The centre of the pattern is called the **centre of rotation symmetry**.

A tracing of this shape will fit over the shape in only one position.
Strictly speaking, the shape has rotation symmetry of order 1, but there is no single centre of rotation symmetry.
We usually say the shape has **no rotation symmetry**.

B1 You need sheet 117.

Write down the order of rotation symmetry of each design.
Mark the centre of rotation with a dot (unless the order is 1).

C Making designs

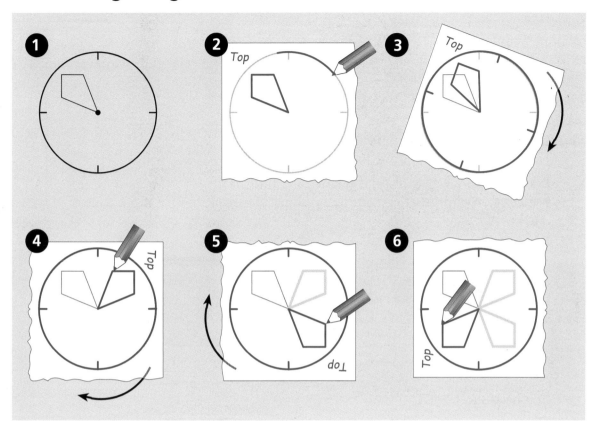

You need sheet 118.

C1 Trace shape B.
Don't forget to trace the circle and guide marks as well.
Use your tracing to make a pattern with rotation symmetry of order 4.

C2 Trace shape C. Make a pattern with rotation symmetry of order 3

C3 (a) Look at shape D on the sheet.
Make a pattern with rotation symmetry of order 4.
Try to do it without tracing paper.
Then check your pattern with tracing paper.

(b) Make patterns with shapes E and F.

C4 Choose a circle on sheet 119.
Draw your own design in the part marked with dotted lines.
Then make a pattern of your own with rotation symmetry.

D Rotation and reflection symmetry

Some patterns and shapes have rotation symmetry **and** reflection symmetry.

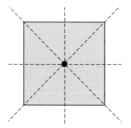

A square has four lines of reflection symmetry and rotation symmetry of order 4.

D1 Copy the shapes below on to square dotty paper. For each shape, draw all the lines of symmetry.

If a shape has a centre of rotation, mark it and write the order of rotation symmetry under the shape.

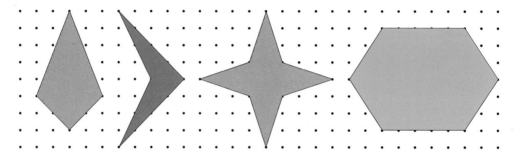

D2 You need sheet 122.

On each shape draw all the lines of symmetry.

If a shape has a centre of rotation, mark it and write the order of rotation symmetry under the design.

D3 Four squares are shaded on this 4 by 4 grid.

(a) What is its order of rotation symmetry?

(b) Does it have reflection symmetry?

D4 On squared paper, draw eight 4 by 4 grids. On each, shade four squares to make a pattern with rotation symmetry (of order 2 or more). Make each pattern different.

Write the order of rotation symmetry under each pattern. Draw in any lines of symmetry on each pattern.

Symmetry in pictures

Lots of pictures and designs have symmetry.

See if you can find some pictures in newspapers and magazines that have rotation or reflection symmetry.

What progress have you made?

Statement

I know when shapes have rotation symmetry or reflection symmetry.

Evidence

1 Which of these shapes have
   (a) rotation symmetry
   (b) reflection symmetry
   (c) rotation symmetry and reflection symmetry

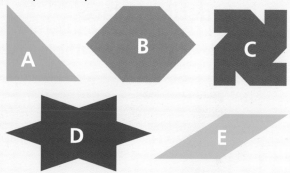

Statement

I can find centres and orders of rotation symmetry.

Evidence

2 Sketch these patterns.
   Mark the centres of rotation and the order of rotation symmetry.

Statement

I can find lines of reflection symmetry.

Evidence

3 On your sketches for question 2, draw the lines of reflection symmetry.

Statement

I can design patterns with rotation symmetry.

Evidence

4 Draw a 4 by 4 grid of squares.
   Shade eight squares in your grid to make a pattern with rotation symmetry of order 4.

5 Decimals 1

This work will help you
- add and subtract numbers with one decimal place
- multiply them by a whole number
- multiply and divide by 10

A One decimal place

A1 Write down the length of each pin, in centimetres.

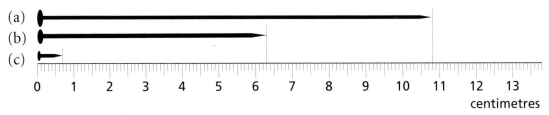

A2 Do the sides of this shape look the same length?

Measure each side in cm. Write down the length of

(a) AB
(b) BC
(c) CD
(d) DA

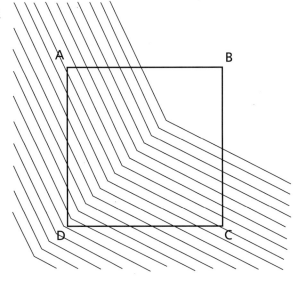

A3 (a) Which of the lines WX, YZ looks longer?
(b) Measure both lines and write down their lengths.
(c) Which is the longer line, WX or YZ?

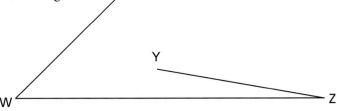

A4 Which of the three lines PQ, RS, TU looks longest? Which shortest?

Measure each line in cm.

Write down the three lengths.

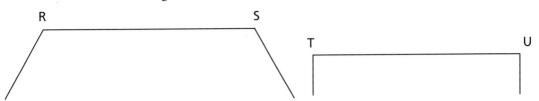

A5 Without measuring, put lines *p*, *q*, *r*, *s* in order of length, shortest first. Now check by measuring. Write down each length.

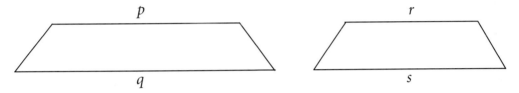

A6 What number does each arrow point to?

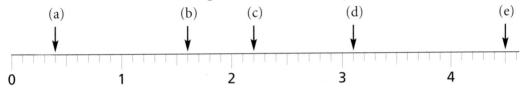

A7 Each jug holds 1 litre when it is full. How much water is there in each jug?

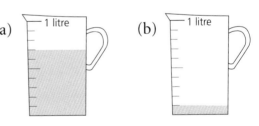

A8 What number does this arrow point to?

A9 Write each list of numbers in order of size, smallest first.

(a) 7, 6.9, 7.2, 6.5, 7.6

(b) 4.2, 6, 5.3, 3.9, 4

(c) 1.2, 0.8, 1.1, 0.6, 1, 2.4

(d) 5, 0.7, 2.8, 3, 2.6, 1.3

B Adding and subtracting

0.7 + 0.5

1.3 − 0.6

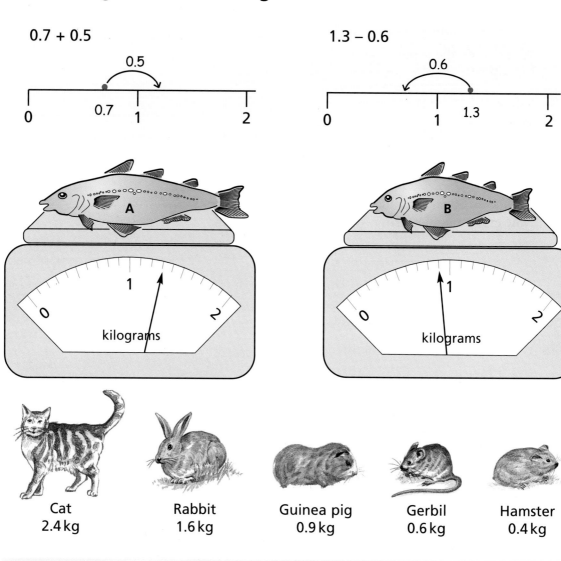

Cat	Rabbit	Guinea pig	Gerbil	Hamster
2.4 kg	1.6 kg	0.9 kg	0.6 kg	0.4 kg

One or two

This is a game for two players.

Put the cards face up on the table.
Take turns to pick up a card.

The winner is the first person to have in their hand
three cards which add up to 1 or 2.
They get one point.

The first person to win 10 points wins the whole game.

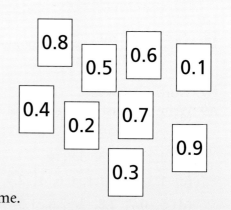

B1 Do these in your head if you can.
 (a) 0.7 + 0.4 (b) 0.9 + 0.3 (c) 1.2 + 0.7 (d) 1.5 + 3 (e) 3.5 + 0.5

B2 Do these in your head if you can.
 (a) 1.6 − 0.2 (b) 1.2 − 0.4 (c) 1.7 − 0.9 (d) 3.7 − 2 (e) 4 − 1.4

B3 Sam and Mel both did 2.3 + 5
Who was right?

```
Sam     2.3       Mel     2.3
      +   5             + 5.0
      ─────             ─────
        2.8               7.3
```

B4 Sam and Mel both did 2.6 + 3.5
Who was right?

```
Sam     2.6       Mel     2.6
      + 3.5             + 3.5
      ─────             ─────
        6.1              5.11
         1
```

Do the questions below either in your head or on paper.

B5

Marrow
3.8 kg

Swede
5.4 kg

Pumpkin
8.0 kg

 (a) How much do the swede and the pumpkin weigh together?
 (b) How much do the marrow and the swede weigh together?
 (c) How much more does the swede weigh than the marrow?
 (d) How much more does the pumpkin weigh than the swede?
 (e) How much more does the pumpkin weigh than the marrow?

B6 Work these out.
 (a) 4.6 + 8.2 (b) 4.7 + 5.6 (c) 7.3 + 0.9 (d) 7 + 4.5 (e) 8.4 + 6

B7 Work these out.
 (a) 4.5 − 1.2 (b) 6.3 − 1.8 (c) 8.4 − 3 (d) 9 − 3.7 (e) 13 − 7.2

B8 Work these out.
 (a) 3.1 − 0.2 (b) 4.5 + 1.5 (c) 7.3 + 3 (d) 6.1 − 2.3 (e) 20 − 4.5

C Multiplying by a whole number

You know that $4 \times 3 = 12$.

This diagram shows that $\mathbf{0.4 \times 3 = 1.2}$

$$\begin{array}{r} 0.4 \\ \times\ 3 \\ \hline 1.2 \\ \hline \end{array}$$

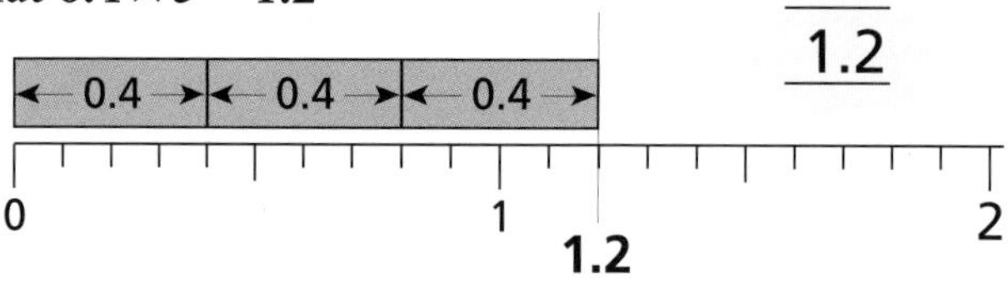

No calculator!

C1 Work these out.

(a) 0.4×2 (b) 0.6×2 (c) 0.5×3

(d) 0.8×2 (e) 0.5×4 (f) 0.4×6

C2 How many litres do these bottles hold altogether?

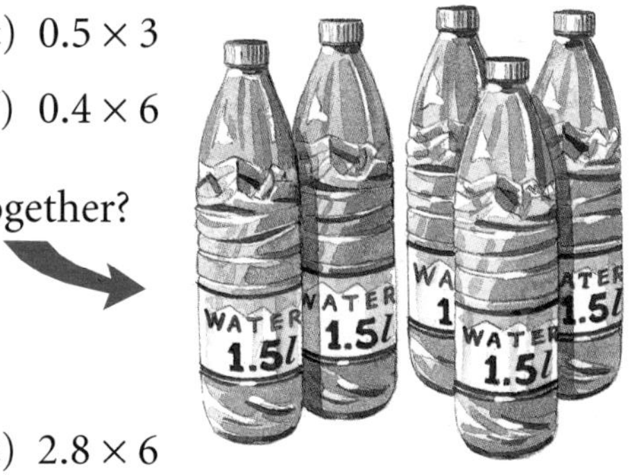

C3 Work these out.

(a) 1.6×6 (b) 4.3×7 (c) 2.8×6

(d) 0.6×7 (e) 2.4×9 (f) 3.7×8

C4 Work out the total weight of each group.

(a)

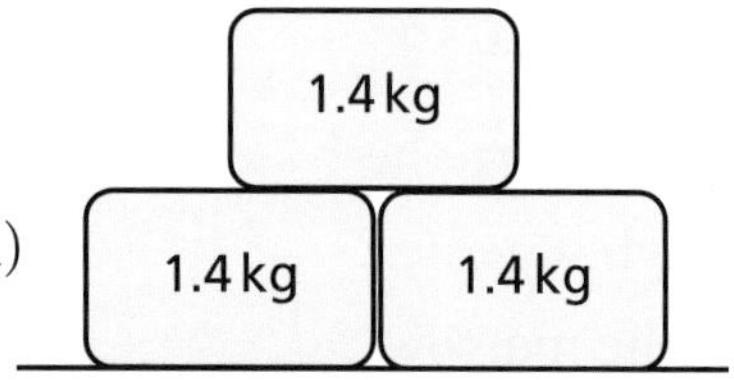

(b)

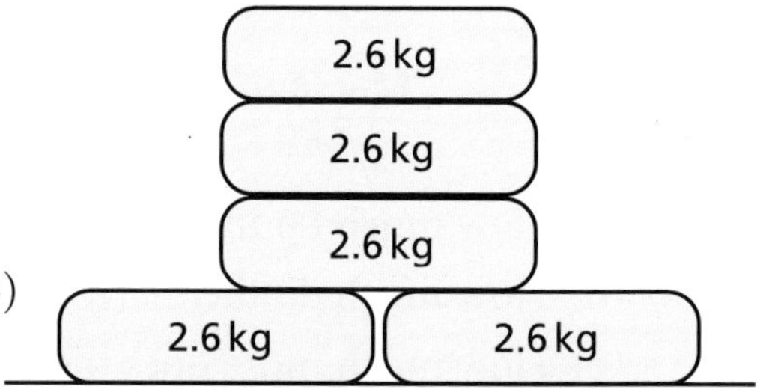

(c)

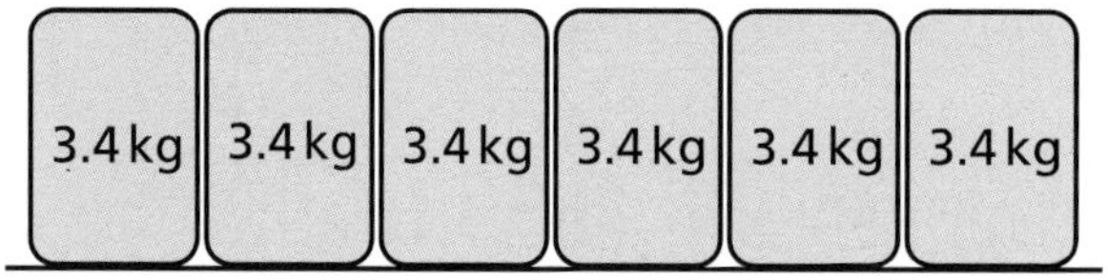

(d)

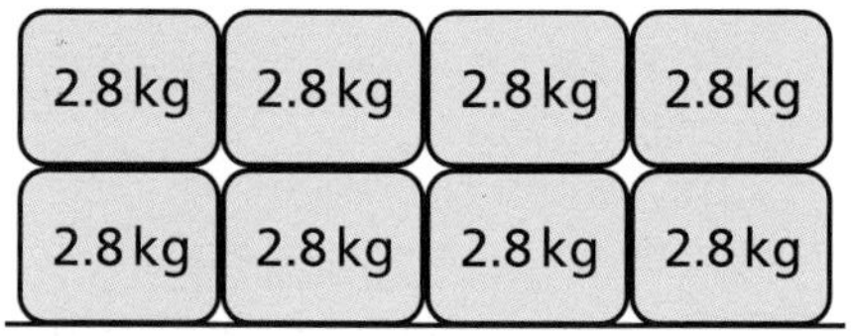

C5 Tina wants to make 8 curtains.
She needs 2.6 metres of material for each curtain.

How much material does she need altogether?

C6 Asad's truck can carry up to 200 kg.
He has 7 boxes, each weighing 28.4 kg.

Can the truck carry all the boxes? Show your working.

D Place value

thousands hundreds tens units . tenths hundredths

1 6 4 5 . 3 2

No calculator!

D1 The figure **3** in the number **1645.3** stands for **3 tenths**, or **0.3**
What do these figures stand for?
(a) the **1** (b) the **6** (c) the **4** (d) the **5**

D2 What do these figures stand for?
(a) the **4** in **647.8** (b) the **9** in **203.9** (c) the **2** in **217.5**

D3 Do these in your head.
(a) Add 1 to 45.6 (b) Add 10 to 45.6 (c) Add 0.1 to 45.6
(d) Add 0.1 to 135.7 (e) Add 1 to 135.7 (f) Add 10 to 135.7

D4 Do these in your head.
(a) 32.8 + 1 (b) 32.8 + 10 (c) 32.8 + 0.1
(d) 137.4 + 10 (e) 137.4 + 0.1 (f) 137.4 + 1

D5 Pam started with this number. **356.4**
She subtracted a number and the result was **306.4**
What number did she subtract?

D6 Dan started with this number. **497.6**
He subtracted a number and the result was **490.6**
What number did he subtract?

D7 Rob started with this number. **685.7**
He subtracted a number and the result was **685.0**
What number did he subtract?

E Multiplying and dividing by 10

Multiplying by 10

- Use a calculator to multiply each of these numbers by 10.

  4.2   2.6   38.7   0.9   1.6   0.2

- What will you get when you multiply each of these numbers by 10? Check each one with a calculator.

  0.6   5.9   72.4   0.8   7.6   0.3

When you multiply by 10,

    tens become **hundreds**

    units become **tens**

    tenths become **units**

So every figure moves **one place to the left**.

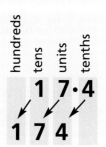

E1 Write down the results of these, without using a calculator.

(a) 3.4 × 10   (b) 0.2 × 10   (c) 1.4 × 10   (d) 2.1 × 10   (e) 0.3 × 10

(f) 10 × 1.7   (g) 10 × 5.4   (h) 10 × 15.1   (i) 2.7 × 10   (j) 10 × 20.4

E2 Do these without a calculator.

(a) 13.7 × 10   (b) 53.2 × 10   (c) 1.8 × 10   (d) 10 × 4.6   (e) 5.6 × 10

(f) 10 × 3.3   (g) 0.7 × 10   (h) 10 × 19.4   (i) 32.5 × 10   (j) 10 × 16.0

Dividing by 10

- Use a calculator to divide each of these numbers by 10.

  37   56   187   8   45   3

- What will you get when you divide each of these numbers by 10? Check each one with a calculator.

  246   12   5   74   2   68

When you divide by 10,

    hundreds become **tens**

    tens become **units**

    units become **tenths**

So every figure moves **one place to the right**.

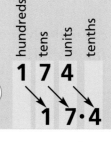

E3 Write down the results of these, without using a calculator.
(a) 43 ÷ 10   (b) 27 ÷ 10   (c) 133 ÷ 10   (d) 648 ÷ 10   (e) 6 ÷ 10
(f) 67 ÷ 10   (g) 25 ÷ 10   (h) 4 ÷ 10   (i) 42 ÷ 10   (j) 125 ÷ 10

E4 Do these without a calculator.
(a) 36 ÷ 10   (b) 9 ÷ 10   (c) 16 ÷ 10   (d) 70 ÷ 10   (e) 31 ÷ 10
(f) 136 ÷ 10   (g) 55 ÷ 10   (h) 130 ÷ 10   (i) 21 ÷ 10   (j) 7 ÷ 10

E5 Do these without a calculator. Some are × 10 and some are ÷ 10.
(a) 4.7 × 10   (b) 36 ÷ 10   (c) 38 × 10   (d) 10 × 0.1   (e) 268 ÷ 10
(f) 157 ÷ 10   (g) 38.6 × 10   (h) 74 ÷ 10   (i) 71 × 10   (j) 19 ÷ 10

E6 Do these without a calculator.
(a) 45 ÷ 10   (b) 10 × 67   (c) 220 ÷ 10   (d) 603 ÷ 10   (e) 10 × 0.2
(f) 65 × 10   (g) 10 × 3.2   (h) 0.4 × 10   (i) 121 ÷ 10   (j) 10 × 22.3

What progress have you made?

Statement	Evidence
I can add and subtract numbers with one decimal place.	1 Work these out. (a) 2.4 + 3.2 (b) 6.7 − 1.4 (c) 2.7 + 0.6 (d) 5.3 + 6 (e) 7.5 − 0.7 (f) 4.5 − 1.6
I can multiply tenths by a whole number.	2 Work these out. (a) 0.6 × 5 (b) 2.7 × 4
I know what the figures stand for in a number with one decimal place.	3 What do these figures stand for? (a) the 7 in 276.4 (b) the 3 in 49.3
I can multiply and divide by 10 without a calculator.	4 Work these out without a calculator. (a) 5.6 × 10 (b) 362 ÷ 10 (c) 231 ÷ 10 (d) 10 × 4.3 (e) 8 ÷ 10 (f) 0.7 × 10

Review 1

1. We can write 'half past eight in the evening' as 8:30 p.m.
   Write these times using a.m. or p.m.
   (a) half past 8 in the morning
   (b) a quarter past 2 in the morning
   (c) a quarter to 6 in the morning
   (d) a quarter to 10 at night
   (e) 20 minutes to 11 at night
   (f) 25 to 7 in the morning
   (g) five to 8 in the evening
   (h) five past 3 in the morning

2. Using a.m. or p.m. write the time that each of these clocks shows.

   (a)
   afternoon

   (b)
   night

   (c)
   morning

   (d)
   evening

3. Miguel is a gardener at Wrotton College.
   He gets to work at a quarter to 9.
   (a) Miguel is asked to mow the front lawn of the college.
       He starts mowing at 5 past 9.
       How long is that after he gets to work?
   (b) Miguel finishes mowing the lawn at 10 minutes to 11.
       How long does it take him to mow the lawn?
   (c) Miguel's tea-break is from 10 minutes to 11 until quarter past 11.
       How many minutes does he get for his tea-break?
   (d) Miguel starts weeding the front borders at a quarter past 11.
       It takes him an hour and a quarter to weed the borders.
       At what time does he finish?

4. What is the probability that red will win on each of these spinners?

   (a)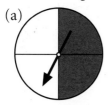
   (b)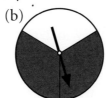
   (c)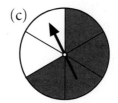
   (d)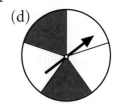

5. What is the probability that red will **not** win
   for each of the spinners in question 4?

6 For each design, say whether it has rotation symmetry.
If it does, write down the order of rotation symmetry.

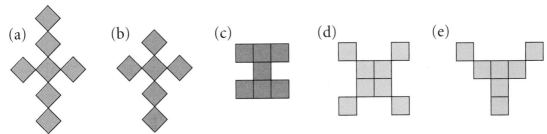

7 Copy each of these grids on to squared paper.
In each copy, shade as few squares as possible so that
the grid has the order of rotation symmetry shown.

(a) Order 2   (b) Order 2   (c) Order 4

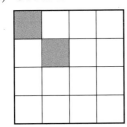 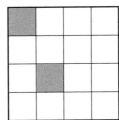 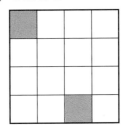

8 What number does each arrow point to?

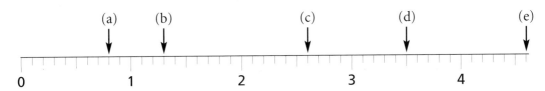

9 Write each list in order of size, smallest first.
(a) 5, 4.9, 0.9, 8, 1.1
(b) 10, 8.7, 9.1, 9.9, 7.5
(c) 0.8, 1, 0.7, 1.1, 0.2
(d) 3.5, 3, 6, 2.9, 5.5

10 Without a calculator, work these out.
(a) 2.3 + 1.2 (b) 4 − 0.6 (c) 3.9 + 0.2 (d) 7 − 1.2 (e) 0.8 + 0.9
(f) 3 − 1.8 (g) 1.9 + 1.9 (h) 6 − 1.4 (i) 6.2 − 1.4 (j) 10 − 5.5

11 Do these without a calculator.
(a) 1.3 × 2 (b) 2.4 × 2 (c) 1.2 × 3 (d) 0.8 × 4 (e) 1.3 × 8
(f) 1.9 × 10 (g) 10 × 2.5 (h) 15 × 10 (i) 10 × 27.3 (j) 18.3 × 10
(k) 55 ÷ 10 (l) 9 ÷ 10 (m) 100 ÷ 10 (n) 17 ÷ 10 (o) 65 ÷ 10

6 Number grids

This is about number grids and algebra.
The work will help you

- solve problems and investigate patterns on a number grid
- simplify expressions

A Square grids

An introductory activity is described in the teacher's guide.

These number grids use the rules '+ 6' across (→) and '+ 2' down (↓).

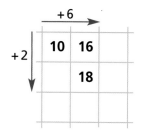
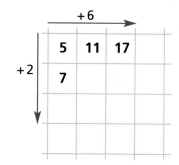

- Copy and complete these grids.

- Draw some different sized square grids that use the rules '+ 6' across and '+ 2' down.

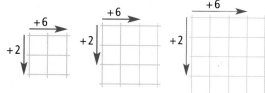

- For each of your grids, choose a different number for the top left corner, Then complete the grid.

- For your grids, find a rule to go diagonally from one number to another.

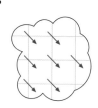

- Use your rule to work out the numbers for the shaded squares in these grids.

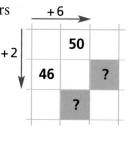
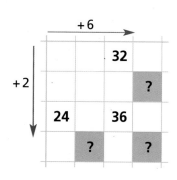

A1 (a) Copy and complete these number grids.

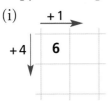   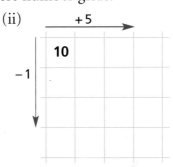   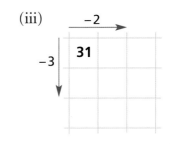

(b) For each grid, find a rule to go diagonally (↘) from one number to another.

A2 Investigate your own number grids.
- Use only '+' or '−' in your rules.
- Choose your own numbers for the top left corner.
- Record your results in a table like the one below.

Across rule	Down rule	Diagonal rule

What is the link between the across, down and diagonal rules?

A3 What are the diagonal rules for these grids?

(a) 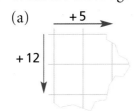   (b)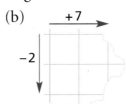

A4 Find some pairs of across and down rules that fit these grids. Compare your rules with someone else's.

(a) 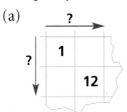   (b)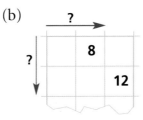

B Grid puzzles

B1 For each puzzle, work out the number in the shaded square.

(a)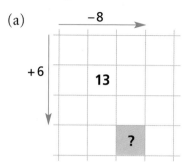

(b)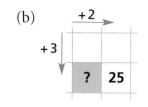

(c)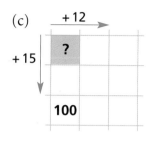

(d)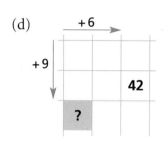

(e)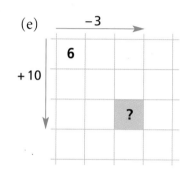

(f)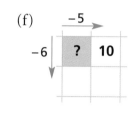

(g)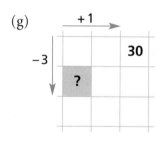

(h)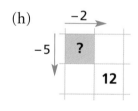

(i)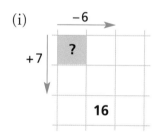

B2 For each puzzle, work out the missing rule.

(a)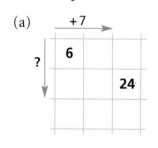

(b)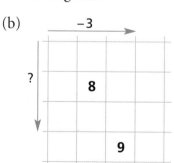

(c)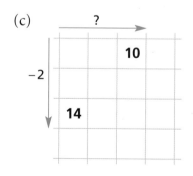

B3 Make up grid puzzles of your own where one of the rules is missing.
Ask someone to solve your puzzles.

B4 Find the across and down rules for each of these number grid puzzles.

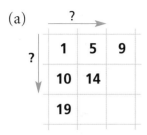
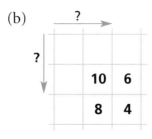
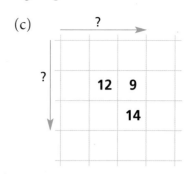

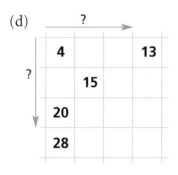
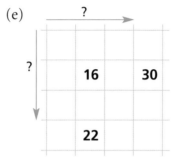

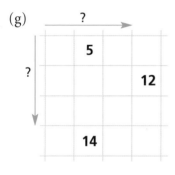
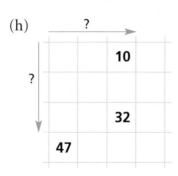
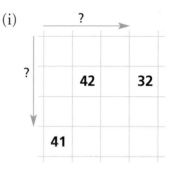

B5 Sam has made up this puzzle. Find four different pairs of rules that fit this grid.

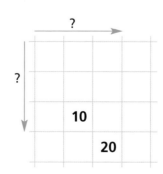

*B6 Ruth is talking about this number grid puzzle.

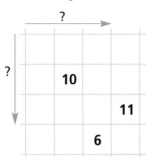

(a) Find the across and down rules for this grid puzzle.

(b) Why do you think Ruth found this puzzle difficult?

*B7 Find the missing rules in these puzzles.

(a) 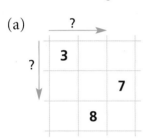   (b) 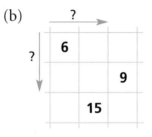   (c)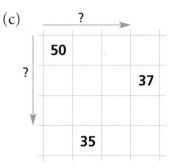

(d) 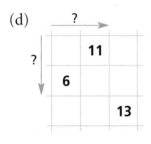   (e)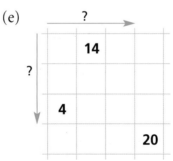

*B8 (a) Find some pairs of rules that fit this grid.

(b) What is the link between the across rule and the down rule each time?

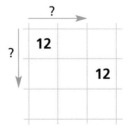

C Grid investigations

C1 Jo makes grids where the across and down rules add and take away the same number.

(a) Copy and complete Jo's grids.

(i) 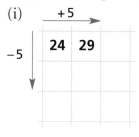 (ii) 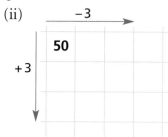 (iii)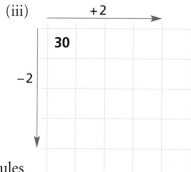

(b) Investigate other grids where the across and down rules add and take away the same number.

(c) What do you notice about these grids?

C2 Suneet uses different numbers in his rules.
He adds the numbers in opposite corners on his grids.
This is one of his grids.

+9

	10	19	28	37
+1	9	18	27	36
	8	17	26	35
	7	16	25	34

Calculations
Add opposite corners
10 + 34 = 44
37 + 7 =

Suneet's calculations

(a) Copy and complete Suneet's calculations.

(b) Add the numbers in opposite corners on these grids.

(i) 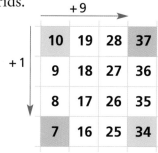 (ii)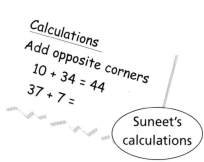

(c) Make some number grids of your own and add the opposite corners.

(d) What do you notice about adding numbers in opposite corners on all these grids?

C3 Sue adds the numbers on each diagonal.
This is one of her grids.

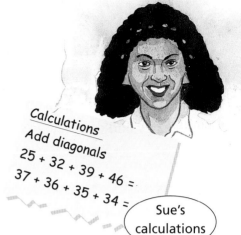

Calculations
Add diagonals
25 + 32 + 39 + 46 =
37 + 36 + 35 + 34 =

Sue's calculations

(a) Copy and complete Sue's calculations.

(b) Add the numbers in each diagonal for these grids.

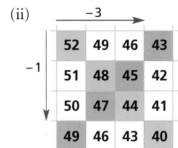

(c) Make some number grids of your own and add the numbers in each diagonal.

(d) What do you notice about adding numbers in diagonals on all these grids?

C4 These grids are all the same size and use the rules '+2' and '−1'.

	+2				+2				+2				+2		
−1	2	4	6	−1	3	5	7	−1	4	6	8	−1	10	12	14
	1	3	5		2	4	6		3	5	7		9	11	13
	0	2	4		1	3	5		2	4	6		8	10	12

(a) Copy and complete the diagonals table.

(b) Draw some more grids like this and find their diagonal totals.

(c) Add your results to the table.

(d) For these grids, find a rule that links the diagonal total and the top left number.

Diagonals table

Top left number	Diagonal total
2	9
3	12
4	
10	

D Algebra on grids

You can use algebra to find rules on grids.
You can write *n* for the number in the top left square and then fill in the other squares.

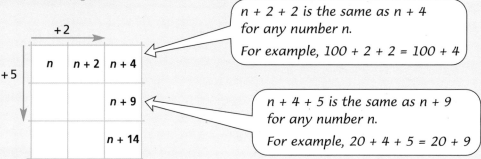

$n + 2 + 2$ is the same as $n + 4$ for any number n.
For example, $100 + 2 + 2 = 100 + 4$

$n + 4 + 5$ is the same as $n + 9$ for any number n.
For example, $20 + 4 + 5 = 20 + 9$

- Copy and complete this grid.
- What is the rule to go from the number in the top left square to the number in the bottom right square?
- Use the rule to work out the number in the bottom right square when the number in the top left square is 10.
- What number in the top left square would give 64 in the bottom right square?
- On the grid, what is a rule to go from the number in the top left square to the number in the bottom left square?

D1

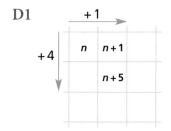

(a) Copy and complete this grid.

(b) Work out the number in the bottom right square when the number in the top left square is 100.

(c) What number in the top left square would give 50 in the bottom right square?

D2 Copy and complete these grids.

(a)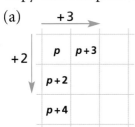
(b)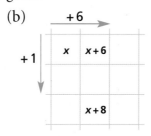
(c)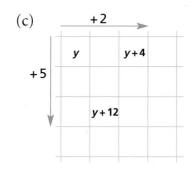

(d)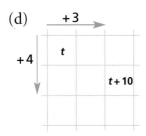
(e)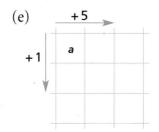

D3 For each grid in question D2, work out the number in the bottom right square when the number in the top left square is 10.

D4 Find rules for these grids.

(a)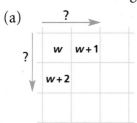
(b)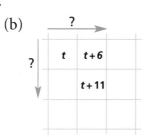
(c)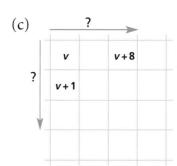

(d)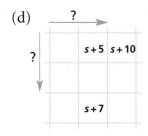
(e)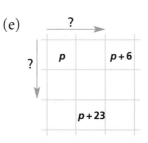

D5 Write each of these in a simpler way.

(a) $f + 3 + 5$
(b) $y + 4 + 4$
(c) $x + 2 + 1$
(d) $z + 5 + 5 + 1$
(e) $p + 4 + 1 + 1$
(f) $m + 2 + 5 + 2$

What progress have you made?

Statement

I can solve number grid problems.

Evidence

1 Copy and complete these grids.

(a) Grid with →+1 across, ↓+3 down; top-left cell 2, below it 5.

(b) Grid with →+5 across, ↓−2 down; top row shows 10, 15.

2 Find the number in each pink square.

(a) →+2, ↓+4; a pink square in the top row, with 10 in the middle row below and to the right.

(b) →−3, ↓+1; a pink square in the top row, with 10 in the bottom row.

3 Work out the missing rules.

(a) →+5, ↓?; 9 in top-right, 6 in middle.

(b) →?, ↓+3; 6 in top-right, 7 in middle.

(c) →?, ↓?; 5 and 17 in top row, 9 in middle.

(d) →?, ↓?; 20 and 15 in top row, 19 in middle.

I can simplify expressions.

4 Copy and complete these grids.

(a) →+1, ↓+6; top row: t, $t+1$; middle row includes $t+7$.

(b) →+?, ↓+?; top row: t, $t+5$; middle row includes $t+8$.

39

7 Stars and angles

This work will help you measure and estimate angles.

A Five-pointed stars

You can draw a five-pointed star with five straight lines.

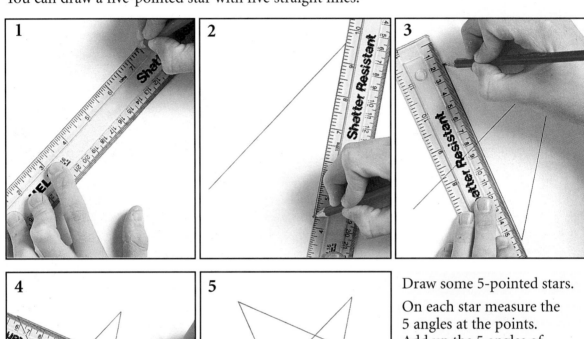

Draw some 5-pointed stars.

On each star measure the 5 angles at the points. Add up the 5 angles of each star.

What do you find?

B Estimating angles

This work is on sheets 130, 131, 132 and 133.

8 Lines at right angles

This work will help you find right angles and draw them.

A Thinking about right angles

A1 The 8 and 5 on this clockface are joined to the centre.

The angle at the centre is a right angle.

What other pairs of numbers make a right angle?

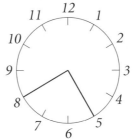

A2 Make a sketch of the eight points of the compass (north, north-east, …).

(a) Jenny is facing west.
She turns through a right angle clockwise.
What direction is she facing now?

(b) Peter is facing south-east.
He turns through a right angle anticlockwise.
What direction is he facing now?

(c) Amal is facing north-west.
She turns through a right angle clockwise.
What direction is she facing now?

B Drawing and checking right angles

A **set square** helps you draw accurate right angles.

Practise drawing some right angles with a set square.

B1 A rectangle has a right angle at each of its four corners.
Follow these instructions to draw a rectangle 9 cm wide by 6 cm high.

1 Draw a line 9 cm long.

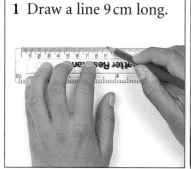

2 At one end, draw a line at right angles.

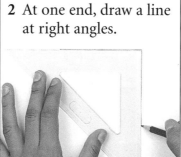

3 Do the same at the other end.

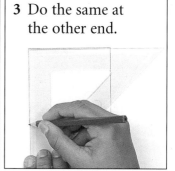

4 Put a mark 6 cm down each of these two lines.

5 Join the marks.

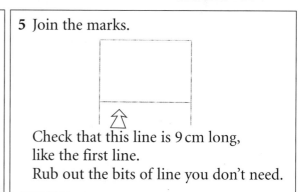

Check that this line is 9 cm long, like the first line.
Rub out the bits of line you don't need.

B2 This is a sketch of a rectangle. Draw the rectangle accurately.

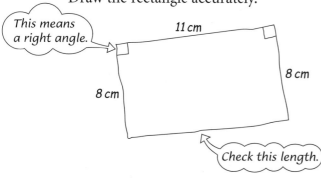

This means a right angle.   11 cm   8 cm   8 cm   *Check this length.*

B3 Draw this shape accurately.

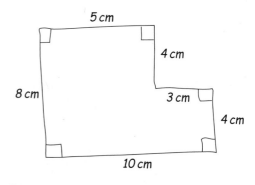

5 cm   4 cm   8 cm   3 cm   4 cm   10 cm

B4 Draw a shape of your own with right angles at all its corners.

B5 Line *a* is at right angles to line *b*.
Check this with a set square.

Some other pairs of lines are at right angles.
Find the pairs.
Try to decide without a set square.
Then check with a set square.
If the lines are far apart, use the corner
of a piece of paper to check.

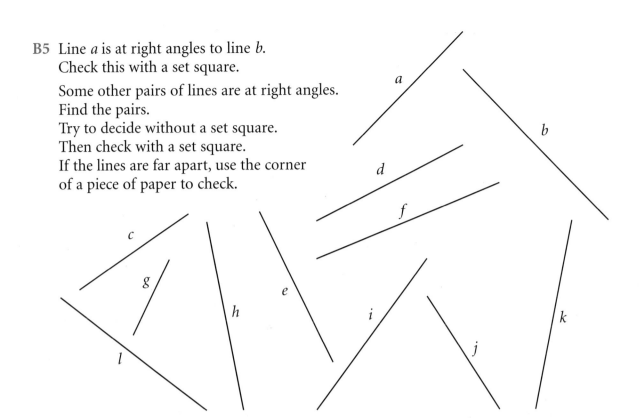

B6 Just by looking, decide which of the corners are right angles.

Now check with a set square.

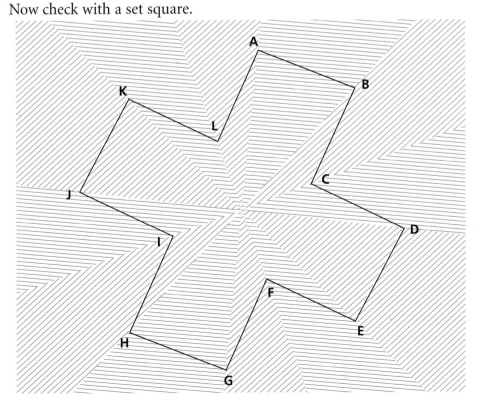

What progress have you made?

Statement

I can find right angles.

Evidence

1 Which lines are at right angles to one another here?

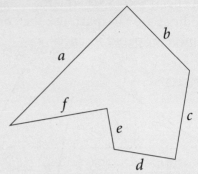

2 Use the corner of a piece of paper to decide which pairs of lines are at right angles.

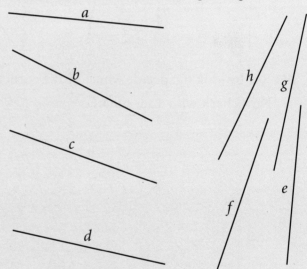

I can draw a shape with right angles accurately.

3 Draw this shape accurately.

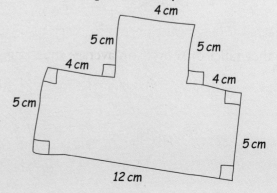

9 Comparisons

This is about ways of comparing groups.
The work will help you
- understand median and range
- use them to make comparisons

A Median

Practical work is described in the teacher's guide.

This school photo was taken in Victorian times.
The pupils on the back row were in 'year 8'.

What is the average height of the seven pupils in the back row?
Are they taller or shorter on average than pupils in your class?

> When the heights are put in order, the middle height is called the **median height** of the group.

A1 Here are the heights of the members of a netball team.

   163 cm   147 cm   154 cm   180 cm   164 cm   156 cm   169 cm

(a) Write them in order, smallest first.

(b) Write down the median height of the team.

A2 (a) Write this group of weights in order, smallest first.

   29 kg   50 kg   44 kg   23 kg   37 kg

(b) What is the median weight of the group?

A3 Find the median weight of each group here.

(a) 76 kg   57 kg   66 kg   52 kg   58 kg   61 kg   70 kg

(b) 89 kg   103 kg   96 kg   94 kg   107 kg

(c) 120 kg   128 kg   117 kg   134 kg   145 kg   126 kg   130 kg   148 kg   124 kg

38 kg comes twice in this group. → 36 kg   38 kg   32 kg   29 kg   38 kg   41 kg   23 kg

You write it twice when you write them in order. → 23 kg   29 kg   32 kg   36 kg   38 kg   38 kg   41 kg

The median weight is 36 kg.

A4 Here are the ages of seven people on holiday.

   23   34   17   13   25   13   16

(a) Write the ages in order, youngest first.

(b) Write down the median age of the group.

A5 Find the median age of each of these groups.

(a) 14   21   16   12   19   14   23   25   13

(b) 20   15   24   11   9   12   17   15   18   20   14

(c) 16   14   26   31   23   14   11   16   20

A6 Here are the lengths of some snakes.

   5.5 m   6.2 m   7 m   4.7 m   7.2 m   6.6 m   6 m   5.2 m   6.8 m

Find the median length of the snakes.

B Dot plots

Here are the heights of a group of girls.

128 cm   139 cm   131 cm   105 cm   153 cm   147 cm   137 cm

They are marked on a **dot plot**.

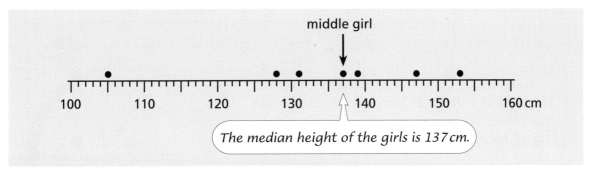

B1 What is the median height of each of these groups?

(a)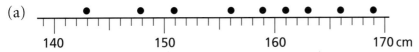

(b)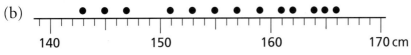

(c)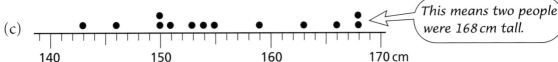

This means two people were 168 cm tall.

(d)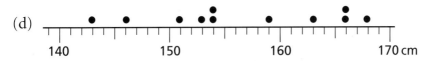

B2 You need sheet 140. Complete part A on this sheet.

B3 A group of pupils write down their heights as

58 cm   148 cm   152 cm   156 cm   160 cm   163 cm   169 cm

(a) What is the median of these heights?

(b) They realise that the pupil who wrote 58 cm was wrong.
Simon said it was probably measured in inches, so 58 cm should be 147 cm.
If this is true, what should the median be?

(c) They then discover that 58 cm should have been written as 158 cm.
What is the correct median height of the group?

C Middle pairs

This dot plot shows the heights of six boys.

There is no 'middle boy',
but there is a middle **pair**.

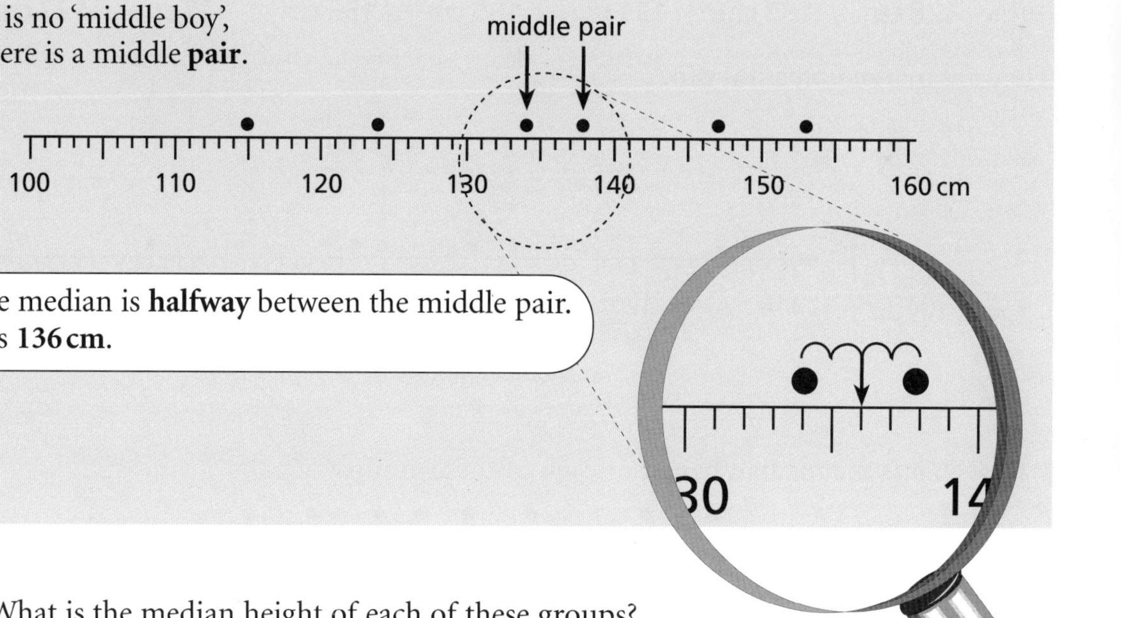

The median is **halfway** between the middle pair.
It is **136 cm.**

C1 What is the median height of each of these groups?

(a)

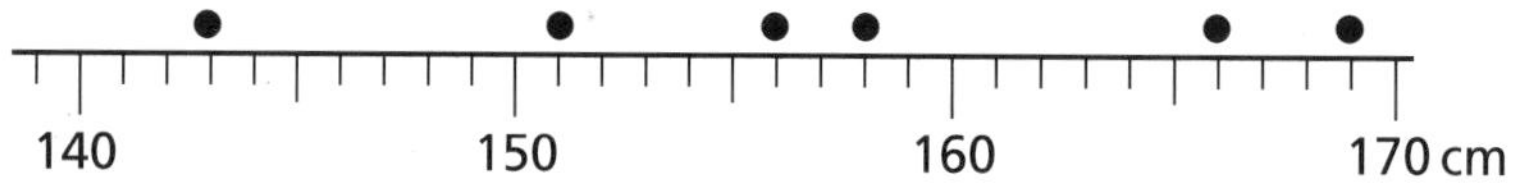

(b)

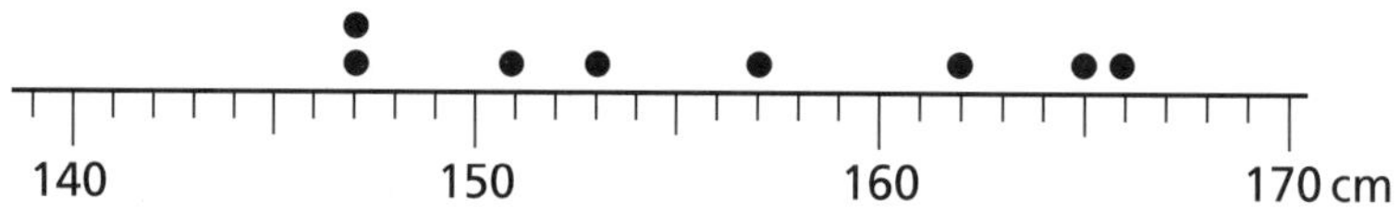

(c)

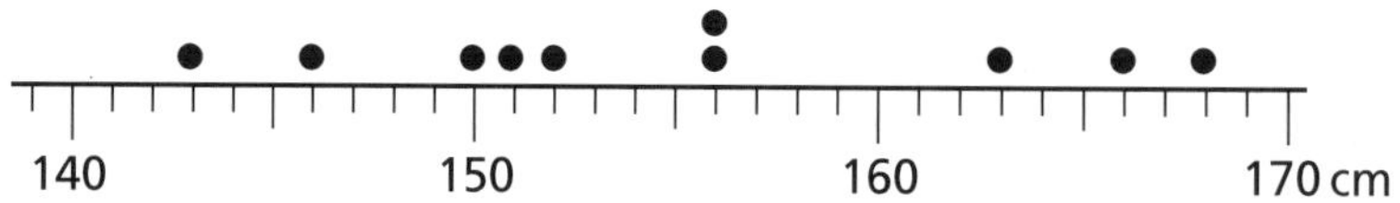

C2 You need sheet 140.
Complete part B on this sheet.

C3 Find the median of each group of heights.

(a) 146 cm 139 cm 140 cm 132 cm 149 cm 153 cm

(b) 174 cm 166 cm 130 cm 155 cm 142 cm 144 cm 147 cm 149 cm

D Comparing groups

For class or group discussion

- Do you think the girls are taller in this picture or the boys?
- What other differences are there in the boys' and girls' heights?

D1 (a) Write down the median height of the girls in the dot plot below.
 (b) Write down the median height of the boys.
 (c) Which group has the greater median height?

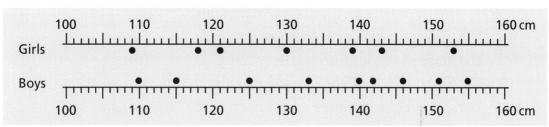

We use the **range** to measure the spread.
The range is the **difference between the largest and smallest**.

D2 (a) Work out the range of each group of heights shown here.

(b) Which group has the greater range?

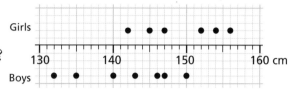

D3 Pick out the largest and smallest from this group of heights (in cm) and work out the range. ➡ 127 142 133 129 140 138

D4 These worms are drawn full size. Find the range of their lengths.

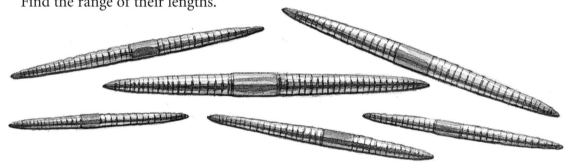

D5 These are the weights, in kg, of two six-a-side football teams.

Team A: 68 72 59 68 77 74   **Team B:** 73 80 65 57 73 74

(a) Find the median weight of each team.

(b) Which team has the greater median weight?

(c) Find the range of weights for each team.

(d) Which team has the greatest range?

D6 These are the weights of some newborn baby boys and girls.

Boys 3.2 kg 2.5 kg 1.9 kg 3.0 kg 2.7 kg 2.4 kg 2.6 kg
Girls 2.9 kg 1.8 kg 3.1 kg 2.5 kg 2.9 kg 3.3 kg 2.0 kg

Write 'True' or 'False' for each of these statements.

(a) The median weight of the group of boys is 2.6 kg.

(b) The range of weights for the girls is 2.9 kg − 2.0 kg = 0.9 kg.

(c) The boys' median weight is higher than the girls' median weight.

(d) The girls' weights are more spread out than the boys' weights.

E How fast do you react?

For working in pairs

Each pair needs a strip from sheet 141.

Cut out the scale from the sheet and stick it on to a ruler (or other wooden or plastic strip).
This makes a reaction timer.

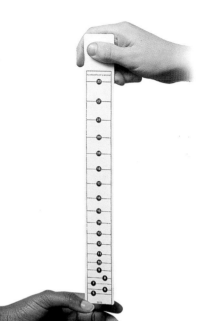

This is how you measure your reaction time.
- Your partner holds the ruler at the top.
- You get ready with your thumb and finger at the zero mark.
- Your partner drops the ruler and you grip it as fast as you can. The scale will tell you your reaction time in hundredths of a second.

Record the results of ten trials each.

You could use a dot plot to record your results, like this.

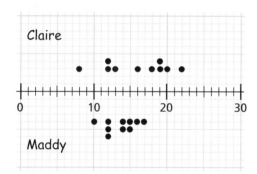

Compare your results with those of your partner using **median** and **range**.

You could also compare your own individual reaction times using your left hand and your right hand.

For the whole class
- Collect either the fastest or the median reaction time for every pupil.
- Compare the reaction times of 12-year-olds and 13-year-olds.

F The Argon Factor

For the class working together

In the mental agility test you are given one minute to memorise the shapes and numbers. 15 questions will be read out to you and you have 5 seconds to write down each answer. You will be scored out of 15.

In the memory test you will be given two minutes to remember the pictures and details of the four people. You will then be given 10 minutes to answer 20 questions on paper to see what you remember.

Do people do better in the tests at the start of the day than at the end of the day?

Are young people's memories better than older people's?

What progress have you made?

Statement	Evidence

I can find the median of a set of data.	1 Find the median of these people's ages.
	  35 23 29 38 42 37 30
	
	2 Find the median of these ages.
	  11 9 6 13 15 19 9 20

I can find the range of a set of data.	3 Find the range of the ages in question 1. 4 Find the range of the ages in question 2.
I can use the median and range to compare two sets of data.	5 Sunil collected his reaction times for each hand. The times (in hundredths of a second) were Left hand 18 9 20 16 23 8 22 24 15 Right hand 12 17 14 11 12 19 13 17 18 16 Find the median time, and the range of times, for each hand. Compare the times for each hand.

10 Ice cream

This work will help you
- practise your arithmetic
- practise your organisational skills

This activity is described in the teacher's guide.

You need sheet 146.

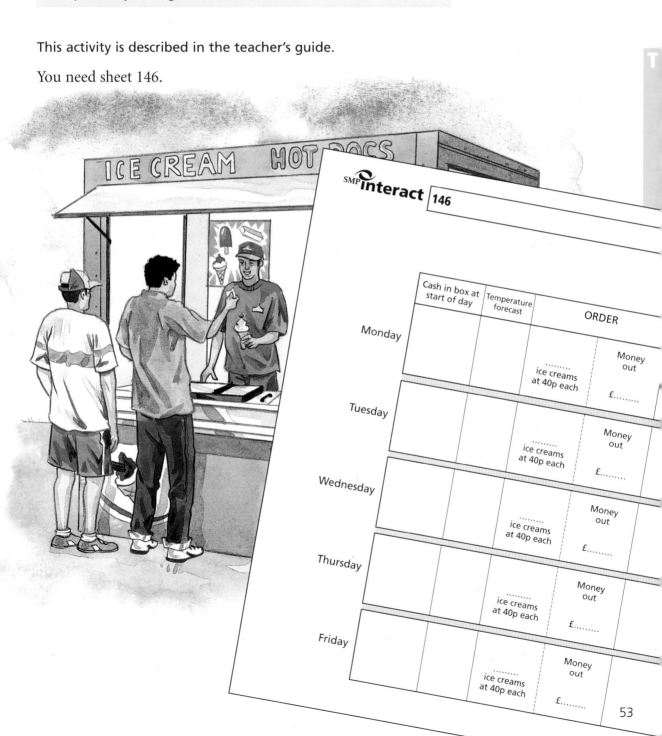

11 Parallel lines

This work will help you draw and find parallel lines.

A Looking for parallel lines

A1 Look at line *a*.
One of the other lines is parallel to it.
Which one is it?

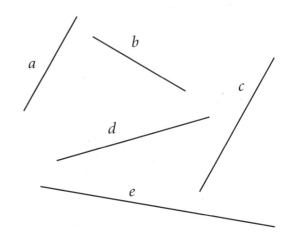

A2 (a) Which of the lines *v*, *w*, *x* or *y* is parallel to line *u*?

(b) There is another pair of parallel lines in the diagram. Which lines are they?

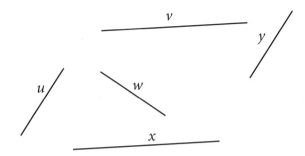

A3 True or false?

(a) A square has two pairs of parallel sides.

(b) A triangle can have one pair of parallel sides.

(c) A hexagon can have three pairs of parallel sides.

A4 There are three pairs of parallel lines in this shape. Which are they?

A5 Which of the lines *b*, *c* or *d* is parallel to line *a*? How can you be sure?

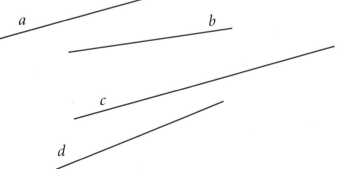

55

B Drawing parallel lines

1 Put a mark on the edge of a piece of paper near its corner.

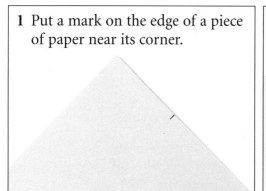

2 Draw a straight line and put the piece of paper along it. Draw against the first mark.

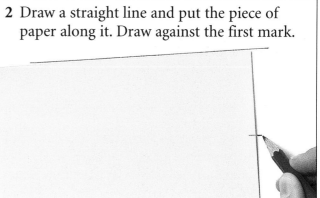

3 Slide the piece of paper along the line. Draw against the mark again.

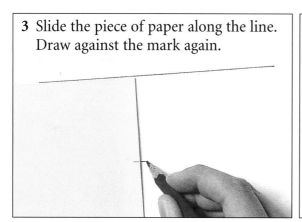

4 Join the two marks with a straight line.

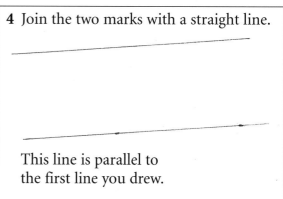

This line is parallel to the first line you drew.

You can make your lines 5 cm apart, for example, by using a ruler.

Measure 5 cm from the edge of the paper.

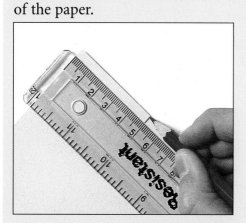

Then do the same as you did above.

B1 Answer the questions on sheet 152.

You can get a set of parallel lines by making several marks on your piece of paper.

This method is good for making designs like the ones below.

You can draw some like them or make parallel line designs of your own.

Work on plain paper.

Use a pencil so you can rub out some lines if you need to.

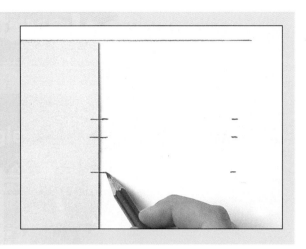

Start by drawing the outside four-sided shape.

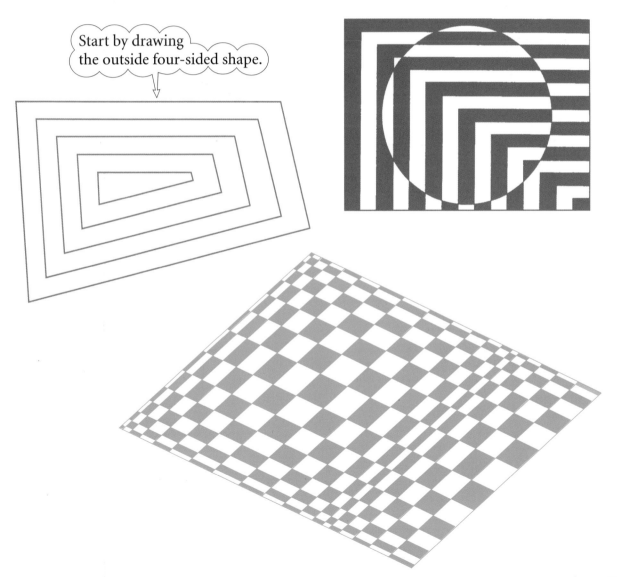

C Checking whether lines are parallel

You can use the corner of a piece of paper to check that two lines are parallel.

1 Put a piece of paper against one of the lines.
Mark where the other line comes on the paper.

2 Slide the edge of the paper along the line it is touching.

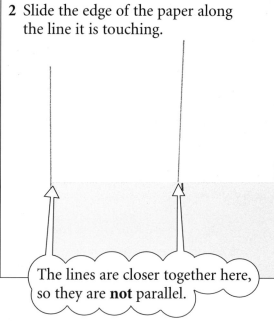

The lines are closer together here, so they are **not** parallel.

C1 Do the questions on sheet 153.

C2 Are the two black lines parallel?

What progress have you made?

Statement	Evidence
I can draw parallel lines.	1 Draw two parallel lines 6 cm apart.
I can check whether lines are parallel.	2 Check whether these two lines are parallel.

12 Anamorphs

This will remind you of work you have done with coordinates.

You need sheets 154 and 155.

Here are the coordinates for four pictures.
Choose a grid from the sheets to draw each one.

Join up each point as you go along.
A ♦ means start a new part of the picture.

Picture 1 ♦ (1, 0) (2, 3) (2, 6) (1, 9) (3, 8) (4, 6) (7, 7) (9, 5) (8, 4) (9, 3) (7, 2) (4, 3) (3, 1) (1, 0)
♦ (7, 5) (7, 6) (6, 6) (6, 5) (7, 5) Shade this part.

Picture 2 ♦ (9, 5) (7, 4) (7, 2) (6, 2) (6, 4) (3, 4) (3, 2) (2, 2) (2, 6) (1, 9) (2, 9) (3, 6) (7, 6)
♦ (7, 5) (7, 6) (6, 7) (6, 9) (7, 8) (9, 8) (10, 9) (10, 7) (9, 6) (9, 5) (7, 5)
♦ Put large dots at (8, 5) and (7, 7) and (9, 7).

Picture 3 ♦ (1, 2) (1, 7) (5, 9) (9, 7) (9, 2) (1, 2)
♦ (2, 5) (2, 7) (4, 7) (4, 5) (2, 5) Shade this part.
♦ (6, 5) (6, 7) (8, 7) (8, 5) (6, 5) Shade this part.
♦ (4, 3) (5, 5) (6, 3) (4, 3) Shade this part.
♦ (3, 2) (3, 1) (7, 1) (7, 2)

Picture 4 ♦ (3, 1) (3, 3) (1, 6) (1, 7) (3, 5) (2, 9) (3, 9) (4, 6) (4, 10) (5, 10) (5, 6) (6, 9) (7, 9) (6, 6) (7, 8) (8, 8) (7, 6) (6, 3) (6, 1) (3, 1)

Artists have used this method to draw strange pictures called **anamorphs**.

The one at the top of this page is from a famous painting.
It is in the National Gallery in London.
Can you see a skull?

Look at your book from the side to read this message.

Review 2

1 For each of these number grid puzzles, find
   (i) the missing rules
   (ii) the number in the shaded square

(a)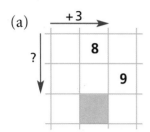

(b)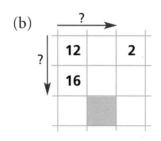

(c)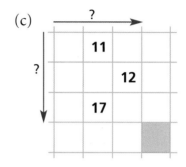

2 Copy and complete these grids.

(a)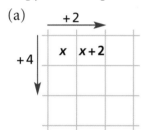

(b)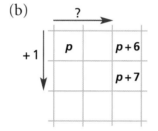

(c)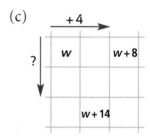

3 Write each of these in a simpler way.
   (a) $z + 2 + 7$
   (b) $k + 4 + 5$
   (c) $m + 1 + 7$

4 Find the median of each of these sets of data.
   (a) The ages of some skaters: 45, 17, 16, 23, 18, 14, 36
   (b) The lengths of some ropes: 4.5 m, 7.3 m, 8 m, 6.6 m, 5.7 m
   (c) The weights of some pupils: 33 kg, 37 kg, 31 kg, 27 kg, 31 kg, 42 kg, 38 kg
   (d) The weekly hours worked by some people at a supermarket:
       36, 52, 44, 36, 40, 42, 39, 48, 30, 47
   (e) The heights of the pupils shown in this dot plot:

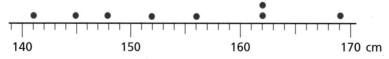

5 These are the pulse rates of a group of boys and a group of girls.

Boys 65 72 88 68 90 75 68 beats per minute

Girls 67 83 77 69 71 79 85 82 beats per minute

(a) Find the median pulse rate for each group.

(b) Find the range of pulses for each group.

(c) Which group had the highest median pulse rate?

(d) Which group had the highest range of pulses?

6 (a) Find three pairs of lines in this diagram which are at right angles.

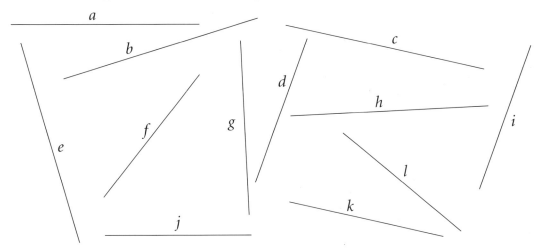

(b) Find three pairs of lines in this diagram which are parallel.

7 Here is a sketch of a shape.
Draw it accurately on plain paper.

You can use the corner of a piece of paper to draw the right angles.

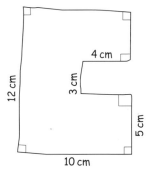

8 Draw a right-angled triangle which is 9 cm long and 5 cm high on plain paper.

Draw four lines, 1 cm apart, parallel to the base.

Draw a line from the top point to the middle of the base.
Colour the pattern as shown here

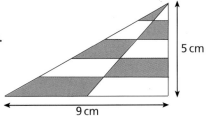

61

13 Practical problems

This work will help you measure and estimate quantities.

Weighty problems for a group of pupils

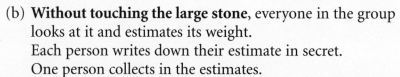

1 There are two stones.

   (a) Someone weighs the small stone.

   (b) **Without touching the large stone**, everyone in the group looks at it and estimates its weight.
   Each person writes down their estimate in secret.
   One person collects in the estimates.

   (c) Now someone weighs the large stone.
   Everyone compares their estimates with the weight of the stone.
   They record everybody's estimates and the real weight.

2 Now everyone looks at the set of objects.

   (a) **Without touching them**, everyone in the group decides what order of weight they should go in.
   Each person writes down their order in secret.

   (b) Now people can touch the objects.
   Without weighing them on the scales, each person writes the objects in order of weight again, in secret.
   One person collects in both sets of estimates.

   (c) Now someone weighs the objects.
   Each person compares their two orders of weight with the real order.

Beans for pupils working individually

Estimate the number of beans in the jar.

Write down how you worked out your estimate and any measurements that you made.

Cornflakes for pupils working individually

1. (a) Get a cereal bowl.
   Put into it enough cornflakes for a reasonable breakfast.
   Weigh this amount.
   (b) Weigh out the suggested portion size given on the box.
   How does this portion compare with your portion?

2. (a) Work out how many portions you can pour from the box if they are the recommended size.
   (b) How many portions can you pour from the box if they are your size?

3. (a) Work out the cost of the recommended portion.
   (b) Work out the cost of your own portion.

Getting better for pupils working individually

You have three bottles. Suppose each one is filled with medicine.

1. Look at these instructions.

   ```
   Take one 5 ml spoonful
         3 times a day.
   ```

   How long will each of the three bottles last?

2. How long will each bottle last with this prescription?

   ```
   Take two 5 ml spoonfuls
         twice a day.
   ```

3. You are told to take two 5 ml spoonfuls 4 times a day for 20 days.
   Would any of your three bottles be big enough?

Children's TV for pupils working individually

You have a friend abroad.
You want to show them what children's TV is like in Britain.

You decide to record a video.
You want to record some typical programmes on a three-hour tape.

Choose some children's programmes to record.
Choose a mixture of different types and different lengths.
Don't just choose your favourite programmes!

Try to use as much of the three-hour tape as possible.

Write down

- the names of the programmes
- how long each one lasts
- the total time of your recording

Windfall for pupils working individually

A rich relative has given you £150 to spend!

You definitely want a watch,
a Walkman and a bag.

You can buy anything else you want.

But you must make the total you spend
as close to £150 as possible.

Decide your shopping list.

Record the catalogue numbers, the prices and the total cost.

14 Angle dominoes

This will remind you of work you have done with angles.

You need sheet 156, cut up into 15 dominoes.

If two angles make up a right angle (90°) they are **complementary**.
If two angles make a straight line (180°) they are **supplementary**.

Domino game

A game for two players

Turn all the dominoes face down together.
Choose five dominoes each and turn them face up.

The player with the highest angle goes first.

Two dominoes with supplementary angles can be placed end to end.

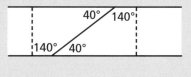

Two dominoes with complementary angles can be placed at right angles.

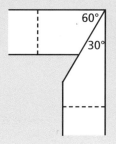

Once you have had your go, you must pick up another domino from the leftovers.

If you cannot go, you just pick up another domino.

The winner is the first to use all their dominoes when there are no more to pick up.

Domino puzzle

A puzzle for one or more people

Can you make a snake using all the dominoes?

The snake can be made only from supplementary or complementary angles.

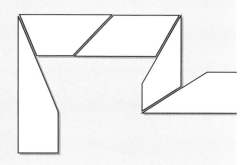

15 Quadrilaterals

This work will help you
- recognise special quadrilaterals (shapes with four straight sides)
- draw them
- learn about their properties

A Special quadrilaterals

A **quadrilateral** is a shape with four straight sides.

What else is special about each of these quadrilaterals?

Lengths of sides?   *Angles?*   *Parallel lines?*   *Reflection symmetry?*

Right angles?   *Rotation symmetry?*

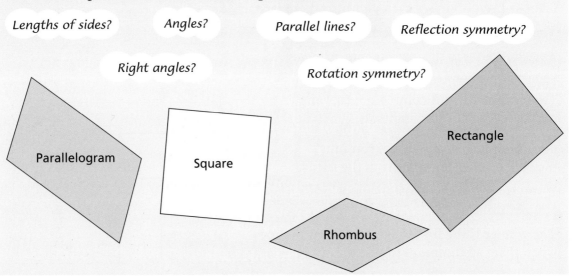

You need sheet 161.

A1 Here is a **parallelogram** drawn on a 5 by 5 grid of spots.

On sheet 161, see how many different parallelograms you can draw with their corners on dots.

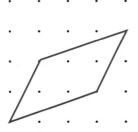

A2 A **rhombus** is a parallelogram with all four sides equal.

On sheet 161, see how many different rhombuses you can draw with their corners on dots.

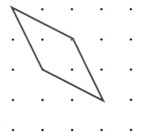

Pop-up parallelogram

Two sheets of A4 paper or thin card can be made into a pop-up mechanism.

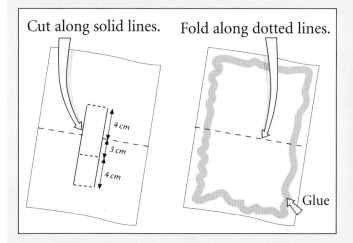
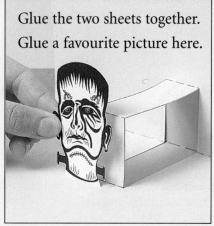

This shape remains a parallelogram as the pop-up is opened.

Other mechanisms, such as this old bridge, have parallelograms in them.

Here are some other types of quadrilaterals.

Trapezium

These sides are parallel.

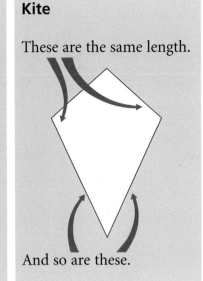

Kite

These are the same length.

And so are these.

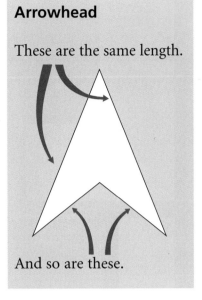

Arrowhead

These are the same length.

And so are these.

You need some copies of sheet 162.

A3 The grid on sheet 162 has lots of rectangles in it. Colour in a rectangle on the first grid.

Try to find as many **different sized** rectangles as you can and shade these on new grids.

A4 Starting on a new row each time, colour in as many different sized quadrilaterals as you can which are

(a) rhombuses (b) kites (c) arrowheads
(d) parallelograms (e) trapeziums

In the middle

On sheet 163 are four quadrilaterals. For each one,
- Write down the name of the original quadrilateral.
- Measure the lengths of the sides in millimetres.
- Divide the lengths by 2 and mark the halfway point along each side.
- Join the midpoints to make a new quadrilateral. Write down its name.
- Repeat these steps on your new quadrilateral, and then once more.
- Write about what you notice.

Repeat this with more quadrilaterals of your own.

B Quadrilaterals from triangles

You will need the triangles from sheet 164.

B1 There are four different types of triangles on sheet 164.
Write down what type of triangle each of A, B, C and D is.

B2 Cut out the two triangles labelled A from the sheet.

(a) Make a kite with the two triangles.
Draw round the kite in your book.
Draw a diagonal to show how it was made.

(b) Make two more quadrilaterals with these two triangles.
These should be different to the kite in (a).
Draw round each quadrilateral in your book.
Label each one with its name.

B3 Cut out the two triangles labelled B.
Show in your book what quadrilaterals you can make with them.

B4 Draw and write in your book to show what quadrilaterals you can make with the two triangles labelled C.

B5 (a) Draw and write in your book to show what quadrilaterals you can make with the two triangles labelled D.

(b) If your D triangles had been shaped like this, 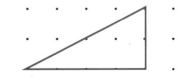 which of the quadrilaterals could you not make?

B6 You can also make some **triangles** using both of the triangles labelled C.
Make some of these triangles and draw round them in your book.

B7 Draw four triangles exactly like this on square dotty paper.
Cut them out.

Use all four of them to make

(a) a square (b) a rectangle
(c) a rhombus (d) a parallelogram
(e) a trapezium (f) a quadrilateral that is not one of these

Draw your answers on square dotty paper and label them with the type of quadrilateral they are.

B8 What fraction of each quadrilateral is shaded?

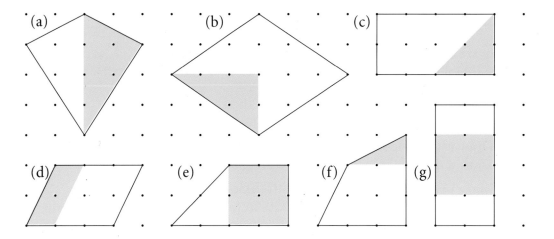

C Accurate drawing

For each of these quadrilaterals,
- Use a ruler and angle measurer to make an accurate drawing.
- Label your drawing with the quadrilateral's special name.
- Think about what the other lengths and angles should be.
- Measure them to see how accurate you were.

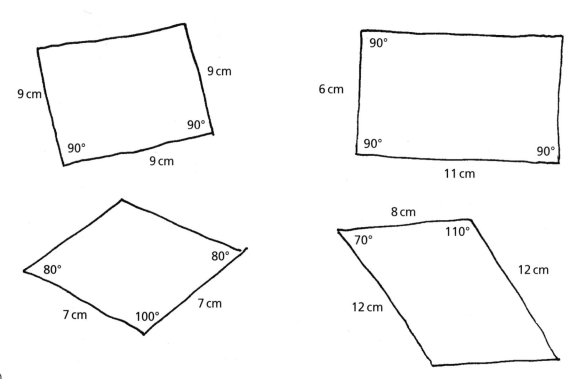

What progress have you made?

Statement

I know the names of special quadrilaterals and their properties.

Evidence

1. Which of these is a trapezium?

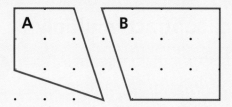

2. Which of these are parallelograms?

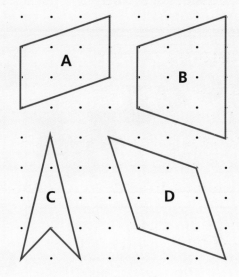

3. What special quadrilaterals have just one line of reflection symmetry?

4. What quadrilateral has four lines of reflection symmetry?

I can draw quadrilaterals.

5. Draw a square with sides 4.5 cm long.

6. Draw a parallelogram from this sketch.

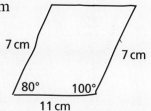

16 Is it an add?

This work will help you decide which calculations to do when you have problems to solve.

A Add, subtract, multiply or divide?

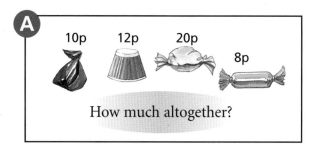

A How much altogether?

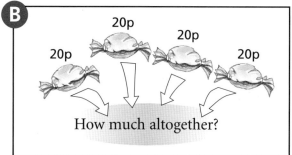

B How much altogether?

C How many less?

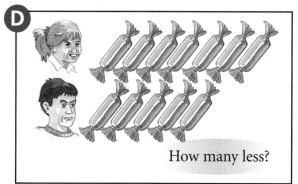

D How many less?

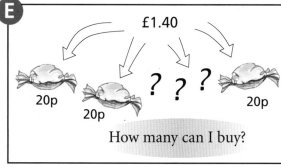

E How many can I buy?

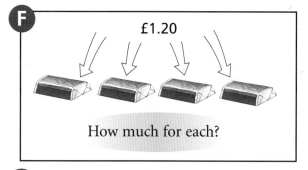

F How much for each?

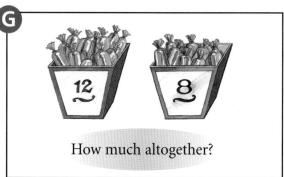

G How much altogether?

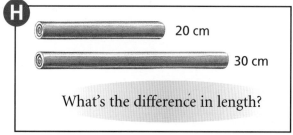

H What's the difference in length?

Calculations

(20 − 4) (20 + 4) (4 ÷ 20) (20 × 4) (4 − 20) (20 ÷ 4)

Bon-bons cost 4p each.

What do 20 bon-bons cost?

How many bon-bons for 20p?

You eat 4 bon-bons.
How many are left?

There is one full box and 4 left in another box.
How many altogether?

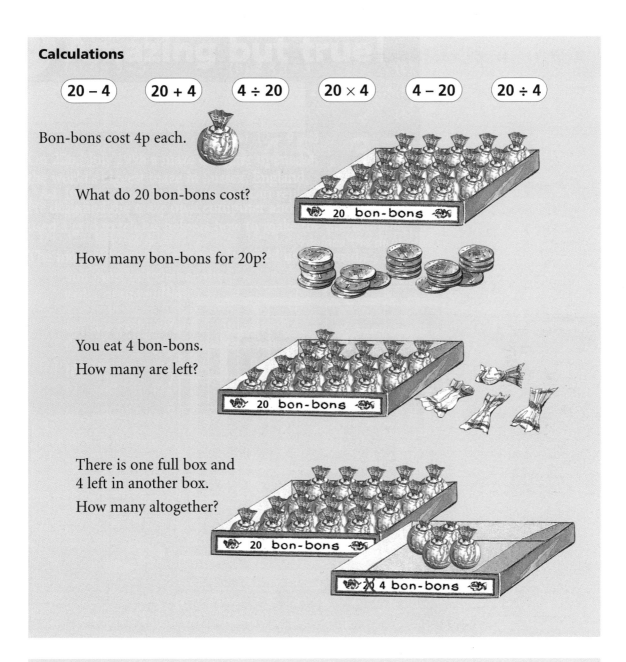

Calculations

(36 − 4) (4 × 36) (4 ÷ 36) (36 × 4) (36 ÷ 4) (4 − 36)

Satya buys some bon-bons.
They cost 36p altogether.
How many does she buy?

Usha buys 36 bon-bons.
What do they cost altogether?

Add, subtract, multiply or divide?

Fit a calculation to each question opposite.

Calculations

nutto
20p

8 nuttos

(8 + 20) (80 × 8) (80 ÷ 20) (80 − 20)
(20 + 80) (80 − 8) (8 × 20) (20 ÷ 8)
(20 × 8) (   ) (80 ÷ 8) (20 − 8)

toffee
8p

12 toffees

(   ) (12 × 8) (20 ÷ 12) (12 − 8)
(12 + 8) (12 ÷ 20) (12 ÷ 8) (20 − 12)
(20 + 12) (8 ÷ 12) (8 × 12) (8 + 12)

cream
12p

Box of creams
£1.20

(12 ÷ 120) (12 − 8) (12 × 8)   (12 × 120)
(12 ÷ 8)   (   ) (12 + 8)   (120 ÷ 8)
(120 − 12) (8 × 12) (120 ÷ 12) (120 × 8)

fudge
30p

20 fudges

Mixed box
(30 of each)

(   )   (20 − 8) (30 ÷ 12) (30 − 12)
(20 × 8)   (   ) (   ) (120 − 30)
(12 ÷ 30) (20 + 8) (120 ÷ 30) (20 ÷ 8)
(120 + 30) (12 − 30) (30 ÷ 120) (12 × 30)

Questions

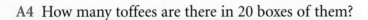

A1 What do 8 nuttos cost?

A2 How many boxes of nuttos can you fill if you have 80 nuttos?

A3 How many nuttos could you buy for 80p?

A4 How many toffees are there in 20 boxes of them?

A5 How much does a box of toffees cost?

A6 How many sweets are there altogether in a box of nuttos and a box of toffees?

A7 How much do you have to pay for 8 creams?

A8 How many creams are there in a box?

A9 How much more do you pay for one cream than one toffee?

A10 You eat 8 fudges from a box of fudges.
How many are left?

A11 What do 12 fudges cost?

A12 You eat 12 creams from a mixed box.
How many creams are left in the mixed box?

A13 The whirls in a mixed box cost £1.20.
How much does one whirl cost?

B Video cassettes

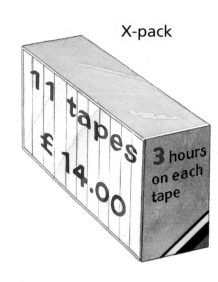

Write the calculation for each question.

B1 How many hours do you get in a W-pack?

B2 What is the cost of one tape from an F-pack?

B3 What would an F-pack and a Y-pack cost together?

B4 What is the difference in price between a Y-pack and an X-pack?

B5 How much would you have to pay for four X-packs?

B6 How many more tapes are there in an X-pack than a Y-pack?

B7 If you buy a W-pack and a Y-pack, how many tapes do you get?

B8 How many tapes are in five Y-packs?

B9 You record a one-and-a-half hour programme on a tape from an F-pack. How much time is left on the tape?

B10 How many tapes from a W-pack would you need to record 15 hours of TV?

C Telling tales

C1 Here is a 'tale' with a question.

> Paul wants to buy 20 cans of Fizzo.
> The cans come in packs of four.
> How many packs does he need to buy?

Which of these calculations goes with the tale?

(20 + 4)   (20 − 4)   (20 ÷ 4)   (4 ÷ 20)

C2 Write down the calculation that goes with each of these tales.

(a) Ellie has 35 soft toys on her bed.
She gives 23 away to a charity
for refugee children.
How many soft toys has she left?

(b) Nasser has 96 tapes.
He wants to store them in racks.
Each rack holds 12 tapes.
How many racks does he need?

(c) Mark has 15 calves in a herd of cows.
One night 8 more calves are born.
How many calves are there now?

(d) Ms King checks brushes for her art lesson.
There are 18 pupils in the class.
She needs five brushes for each pupil.
How many brushes does she need?

C3 Write a tale with a question for each of these calculations.

(a) 6 × 4   (b) 24 ÷ 3   (c) 30 − 12   (d) 12 + 8

What progress have you made?

Statement

I can choose the right calculation for a problem.

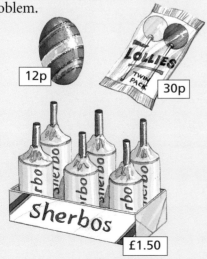

Evidence

1 Write the calculation that goes with each of these questions from this picture.

(a) How much would 6 eggs cost?

(b) What is the cost of one Sherbo in a tray?

(c) How much more does the lollies twin pack cost than an egg?

(d) Angie buys a box of Sherbos and a lollies twin pack.
How much do they cost altogether?

2 Write the calculation for this tale.

> Mary has 6 cats.
> They each eat 4 tins of tuna a week.
> How many tins do they eat altogether in a week?

17 Desk tidy

This work will help you
- measure in centimetres
- make a simple scale drawing

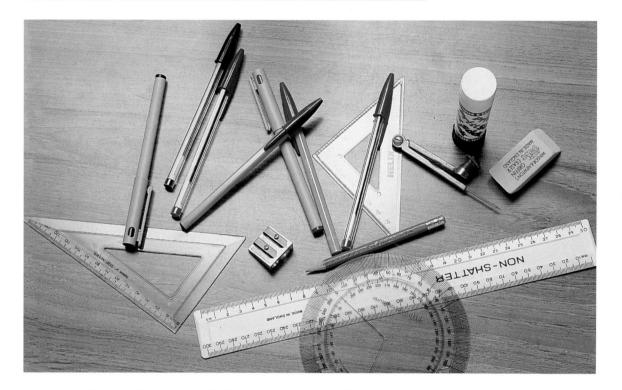

Design a desk tidy

Design a small tray to put on your desk.
It must be big enough to hold all the pens, pencils and other things you keep on your desk.

Make a list of all the things you would keep in the tray.

What are the lengths of your pencils?
How wide are your pencils?

Measure the length and width of the other things.
Add these measurements to your list.

Draw a plan of the tray you would make.
Draw in some of the things to check it is the right size.

You could make one out of card to make sure.

18 Frequency

This work will help you
- understand frequency tables
- read information from frequency bar charts
- draw frequency bar charts
- use frequency bar charts to make comparisons

A Keeping order

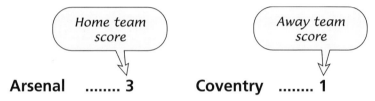

Arsenal 3   **Coventry** 1

The home team is always given first.

This match was a **home win**.

The **total** score in this match was 4 goals.

Is a home team more likely to win than an away team?

You need sheet 165 'English Football League – scores from yesteryear'.

A1 The sheet shows some football results in March 1900, 1930 and 1950.
Look at the results for 1900.

(a) How many of the matches were home wins?

(b) How many of the matches were away wins?

(c) Copy and complete this table for the 1900 games.

March 1900	Tally	Total
Home win	/	
Away win		
Draw	/ /	

A2 (a) Make another table for the 1930 results.
Count up the number of home wins, away wins and draws for March 1930.

(b) Do the same for the March 1950 results.

A3 (a) In the March 1900 results, which type of result was most common?

(b) Which type of result was most common in the March 1930 results?

(c) Which result happened the most often in March 1950?

(d) Look at your own results.
Which type of result was most common?

Another way to say 'how many times' or 'how often' is to say 'how **frequently**'.

The **frequency** of something is the number of times it happens.

This table shows how many times each type of result happened in some 1970 results.
It is called a **frequency table**.

Type of result	Frequency
Home win	13
Away win	9
Draw	7

A4 (a) What was the frequency of draws in the 1900 results?

(b) How frequently did a draw happen in the 1930 results?

(c) What happened more frequently in the 1950 results, a draw or an away win?

B Names, words and letters

How gud is yor spelin?

This activity is described in the teacher's guide

B1 Some pupils were asked to spell the word 'forty'.
They spelt this in several different ways.

The results are shown in this frequency table.

Spelling	Tally	Frequency
fourty	𝍸 𝍸 𝍸 𝍸	20
fourtie	𝍸 𝍸	10
forty	𝍸 𝍸 ////	14
other ways	𝍸	5

(a) How many times was the spelling 'fourtie' used?

(b) How many times was 'forty' spelt correctly?

(c) What was the most frequent wrong spelling of 'forty'?

(d) How many pupils took part in the experiment altogether?

B2 Another word commonly spelt wrongly is 'success'.
This **frequency bar chart** shows the number of times spellings were used in a test.

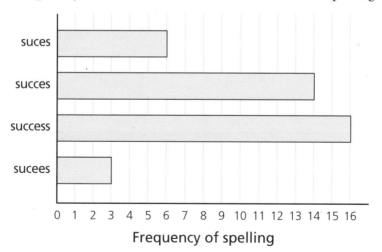

(a) How many times was 'success' spelt correctly?

(b) Which was the most frequent incorrect spelling?

(c) How many times was it spelt incorrectly altogether?

B3 A school adopted a whale in a conservation campaign.
A survey was carried out to find a name for the whale.
This frequency graph shows how pupils voted for some suggestions.

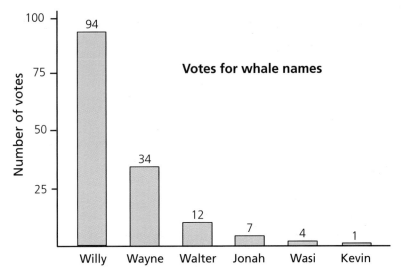

(a) What was the most frequent name voted for?

(b) What information do the numbers on top of the bars give?

(c) What was the frequency of pupils voting for Walter?

(d) What was the frequency of voting for Kevin?

B4 The longest place names in different parts of the British Isles are

Wales: LLANFAIRPWLLGWYNGYLLGOGERYCHWYRNDROBWLL-LLANTYSILIOGOGOGOCH

England: BLAKEHOPEBURNAUGH

Scotland: COIGNAFEUINTERNICH

Ireland: MUCKANAGHEDERDAUHAULIA

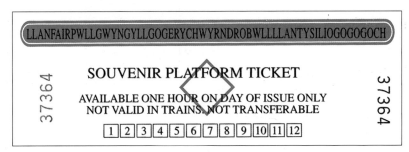

In which country does the longest place name have

(a) the greatest frequency of the letter 'L'

(b) the greatest frequency of vowels (A, E, I, O or U)

C The mode

The item which is most frequent is called the **mode** or **modal value**.

C1 This bar chart shows the frequency of a, e, i, o and u in a piece of English writing.

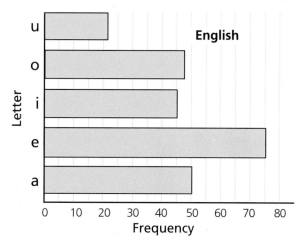

(a) Which letter is the mode on the bar chart?

(b) Which letter occurred 50 times?

(c) Which letter occurred less than 30 times?

(d) Write down the five letters in order of frequency, the mode first.

C2 This chart shows the frequency of a, e, i, o, and u in a piece of French writing.

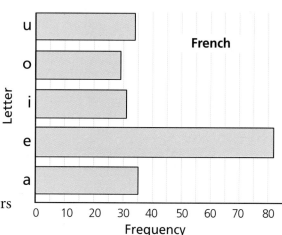

(a) What is the mode of these letters in the French writing?

(b) Say whether these statements are true or false

(i) 'e' is the mode for both the English and the French pieces of writing.

(ii) In the French piece, 'u' is the least frequent letter.

(iii) 'a' was the second most frequent of the letters in the English and French pieces.

D How many times?

A group of pupils have counted the spots on some ladybirds.
Here is a chart showing the results of the survey.

Number of spots on ladybirds

Two pupils are arguing over the results.
Who is correct?

It shows that five ladybirds had six spots.

No way!
It shows six ladybirds had five spots.

D1 (a) How many ladybirds had
   (i) 4 spots   (ii) 7 spots   (iii) 6 spots
 (b) What was the modal number of spots?
 (c) How many ladybirds were used in the survey?
 (d) How many ladybirds had more than 4 spots?

D2 The amount of sky which is covered by clouds is measured in 'oktas'.

| When the sky is completely covered, the cloud cover is 8 oktas. | When the sky is half covered, the sky cover is 4 oktas. |

| When the cloud cover is 0 oktas, the sky is clear. | The cloud cover here is 7 oktas. The sky is nearly all covered. |

Here are the cloud cover figures for a town in September.

```
7  8  8  2  1  2  6  8  7  3
0  0  1  6  7  7  8  3  0  1
1  5  8  4  7  7  6  1  0  3
```

(a) Copy and complete this frequency table.

Cloud cover	Tally	Number of days
0		
1		
2		
3		

(b) Draw a frequency bar chart from this.

(c) What is the modal cloud cover for the town in September?

E Cars and people

This section is for a group to work through together.

Two students are looking at the number of people in cars.

They count the number of people including the driver.
They stand on a motorway bridge to see both ways.

One looks at cars going west. The other looks at cars going east.

Here is a frequency bar chart showing both their results

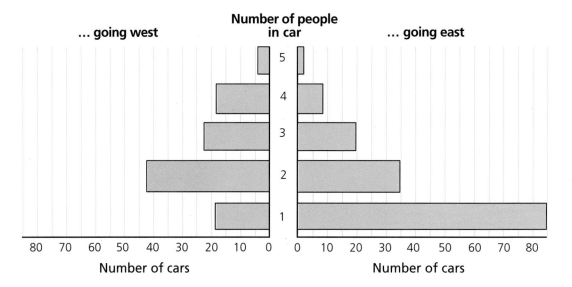

Here are some statements about the chart.
Which do you think are correct?

1 *Of the cars going east, 20 had 3 people in them.*

2 *Of the cars going east 3 had 20 people in them.*

3 *Roughly 42 cars going west had 2 people in them.*

4 *More than half the cars going east had only one person in them.*

5 *Going west the modal number of people in a car was 1.*

6 *Going east the modal number of people in a car was 1.*

7 *There were a lot more cars going east.*

E1 Some pupils carry out a survey on people in cars in a seaside town.
They do this for one hour in July.

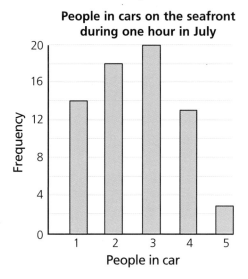

(a) How many of the cars had only the driver and no passengers?

(b) How many cars were counted altogether in the July survey?

E2 Here are the results for a similar survey in December.
This was also recorded for one hour

Here are the raw results.

1	3	1	2	1	1	1	2	1	2
1	2	1	2	3	4	1	1	2	1
2	1	1	2	1	1	2	2	1	1
2	3	3	2	1	1	1	1	2	3
1	1	1	1	1	3	2	2	1	1

(a) Make a frequency table showing these results.

(b) Use your frequency table to draw a frequency bar chart.

(c) Describe briefly what your bar chart shows about the number of people in cars in December.

E3 Compare the bar charts for December and July.
What differences are there?
What might explain the difference?

What progress have you made?

Statement

I can make frequency tables from results.

Evidence

1 A year 8 class were asked how many pieces of fruit they ate last week.

Their answers were
0 1 4 5 3 2 4 1 0 2 2 3 2 2 1 1 1
0 0 3 2 1 1 3 2 1 1 2 1 0

Put this information in a frequency table.

I can read information from frequency tables.

2 (a) How many pupils ate no fruit last week?
  (b) How many people ate more than 2 pieces of fruit?
  (c) How many people were in the class?

I can draw bar charts from a frequency table.

3 Draw a frequency bar chart of these results.

I can find the mode of a set of results.

4 What is the mode of the number of pieces of fruit eaten?

I can compare two sets of results using frequency.

5 A year 10 class were also asked how many pieces of fruit they had eaten.

Here are their results.

Pieces of fruit	Frequency
0	9
1	8
2	5
3	4
4	0
5	4

Write two statements about the differences between the two classes.

⑲ Amazing but true!

This will remind you of work you have done with coordinates.

On 22nd July 1998 a maze designer opened the world's biggest maze in Sussex, England.

He designed the maze on a computer and used a grid to tell his helpers how to make the maze.

The maze below has also been designed using a grid.

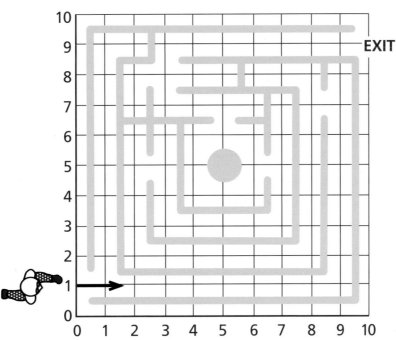

Here is the start of the instructions for getting through the maze.

Start at (0, 1)   Go straight across
At (9, 1)   Go left
At (9, 7)   Go left

Copy and complete the instructions to go through the maze.

Design a maze of your own on a 10 by 10 grid.
Ask a friend to work out the instructions to get through.
Get them to read out their instructions and you check them.

Review 3

1 Here are some quadrilaterals drawn on dotty paper.

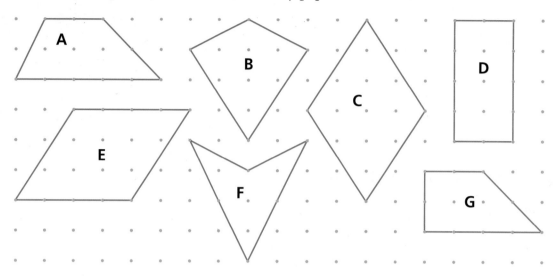

(a) Which of these shapes is a kite?
(b) Which two of these shapes are trapeziums?
(c) Which of these shapes has four right angles?
(d) Which two of these shapes have exactly one line of symmetry?
(e) Which two of these shapes have exactly two lines of symmetry?
(f) Which shape has rotation symmetry order 2 but no reflection symmetry?
(g) Shape G can be cut from the bottom left corner to the upper right corner. This makes two triangles. What type of triangles are these?

2 (a) Use a ruler and angle measurer to make accurate copies of these shapes.

(i) 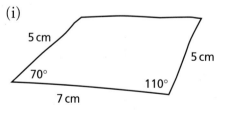   (ii)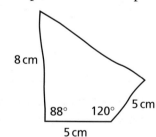

(b) Write the name of each shape inside it.

3 Here are some items at a greengrocer's.

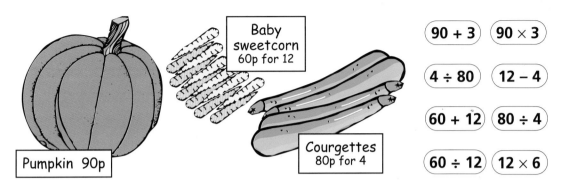

Which of the calculations goes with each of these questions?
(a) What do 3 pumpkins cost?
(b) What is the cost of one courgette?
(c) How many baby sweetcorn are there in 6 packets?
(d) Amanda buys a packet of baby sweetcorn.
She uses 4 of them.
How many does she have left?

4 Class 8J at Grange School were asked how many brothers or sisters they have.
Here are their results.

| 0 | 1 | 3 | 2 | 0 | 4 | 2 | 2 | 2 | 2 | 3 |
| 3 | 2 | 0 | 1 | 1 | 2 | 2 | 1 | 3 | 6 | 1 |

(a) Make a frequency table showing these results.
(b) What is the modal number of brothers and sisters in 8J?
(c) Use your frequency table to draw a bar chart.

5 This bar chart shows the number of brothers and sisters of pupils in class 8M at Waveney School.
(a) How many pupils are there in 8M?
(b) How many pupils had no brother or sister?
(c) What is the modal number of brothers and sisters for this class?
(d) What differences are there between the numbers of brothers and sisters for the two classes?

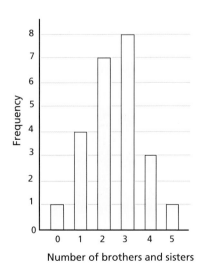

20 Photo display

> This work is about measuring and calculating with measurements.

When you have a photo and a piece of card to stick it on, how can you make sure the photo is in the centre?

1 Cathy wants to make sure the photo is in the centre of the piece of card.

   (a) How wide must the top and bottom borders be?

   (b) How wide must the side borders be?

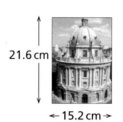

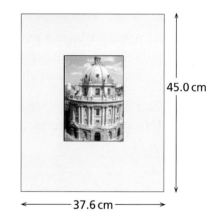

2 Peter wants each photo to be in the middle of the card. Work out the widths of the borders.

(a)

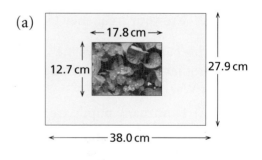

(b)

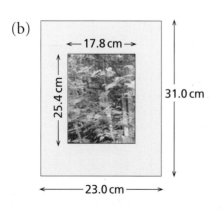

(c)

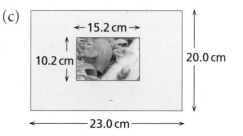

21 Fractions 1

This work will help you recognise fractions of shapes.

A Equal parts?

Part of each of these shapes is red.

- Is the red part a half ($\frac{1}{2}$)?

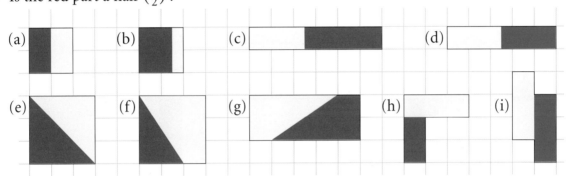

- Is the red part a third ($\frac{1}{3}$)?

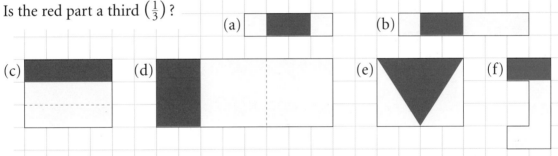

A1 Is the red part $\frac{1}{2}$?

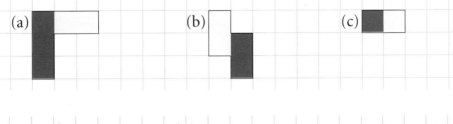

A2 Is it $\frac{1}{3}$?

A3 Is it $\frac{1}{4}$?

A4 Is it $\frac{1}{5}$?

A5 Is it $\frac{1}{6}$?

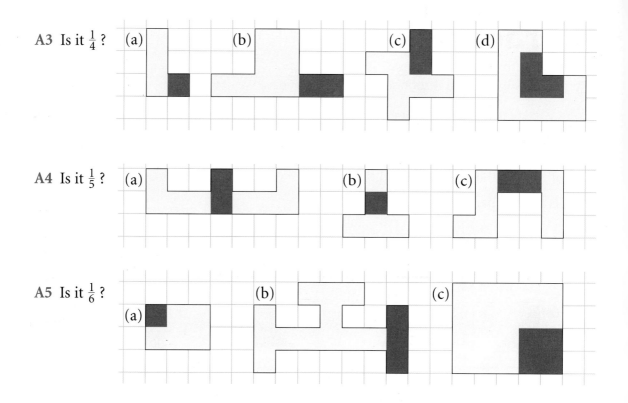

B More than one part

Part of each shape is blue.

- Is the blue part two thirds $\left(\frac{2}{3}\right)$?

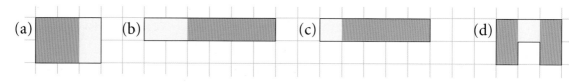

- Is it three quarters $\left(\frac{3}{4}\right)$?

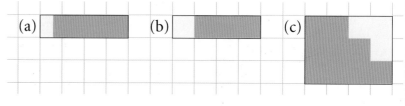

- Is it three fifths $\left(\frac{3}{5}\right)$?

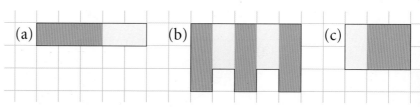

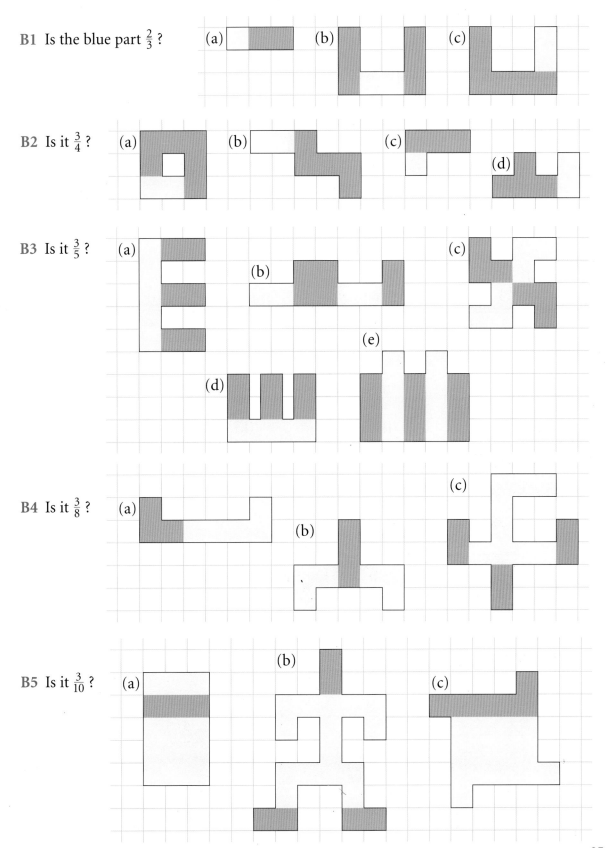

C Coloured tiles

C1 What fraction of each tile is coloured?
What fraction is white?

(a)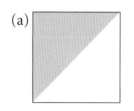
(b)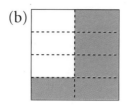
(c)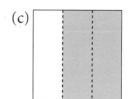
(d)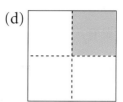

(e)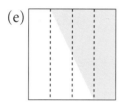
(f)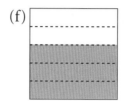
(g)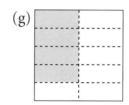
(h)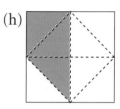

(i)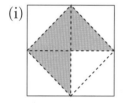
(j)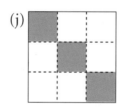
(k)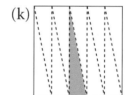
(l)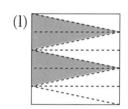

(m)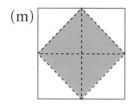
(n)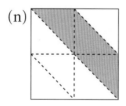
(o)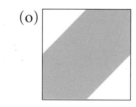
(p)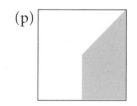

(q)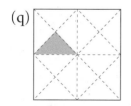
(r)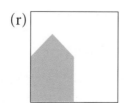
(s)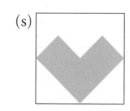
(t)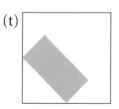

C2 On squared paper, draw 4 copies of this tile.

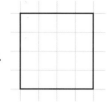

(a) Shade a quarter of the first copy.
(b) On the second copy, shade a quarter in a different way.
(c) Shade $\frac{3}{8}$ of the third copy.
(d) Shade $\frac{3}{8}$ of the fourth copy in a different way.

C3 For each part below draw a tile any size you like and shade

(a) $\frac{1}{2}$ of the tile (b) $\frac{7}{8}$ of the tile (c) $\frac{2}{3}$ of the tile (d) $\frac{1}{5}$ of the tile

C4 Design your own tiles using two colours.
For each tile write the fraction of the tile which is each colour.

C5 This is a tiled floor.
What fraction of the floor is blue?

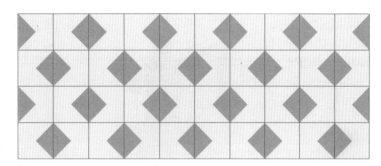

C6 Design your own tiled floor using two colours.
Say what fraction of the floor is each colour.

What progress have you made?

Statement

I can recognise fractions of shapes.

Evidence

1 Is $\frac{1}{3}$ of each shape red?

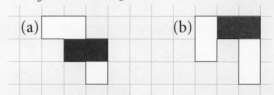

2 What fraction of each shape is red?

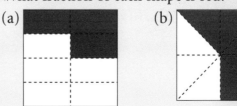

97

22 Enlargement

This is about enlarging shapes.
The work will help you
- spot enlargements
- draw enlargements
- use scale factors for enlargements

A Spotting enlargements

Len is a photographer. He has taken this photo of his brother Bert.

Len makes some enlargements of the photo. He also makes some fakes.

A

D

B

C

F

G

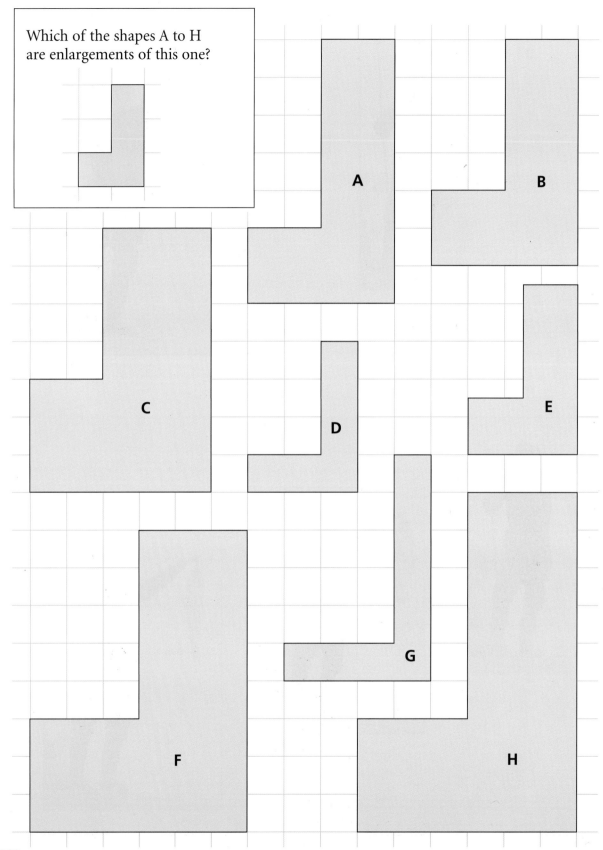

B Enlarging shapes

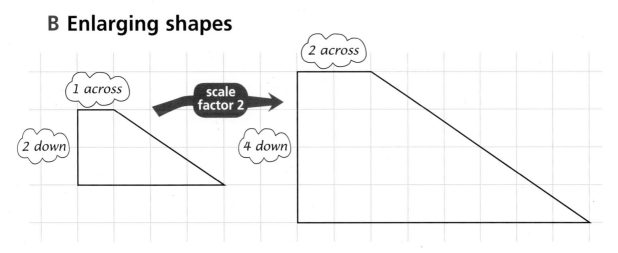

An enlargement which makes all the lengths twice as long as the original is called an enlargement **with scale factor 2**.

An enlargement which makes all the lengths 3 times as long has a scale factor of 3.

B1 This question is on sheet 167.

B2 This question is on sheet 168.

B3 This question is on sheet 169.

B4 Shapes T, U and V are enlargements of shape S.
What is the scale factor of each enlargement?

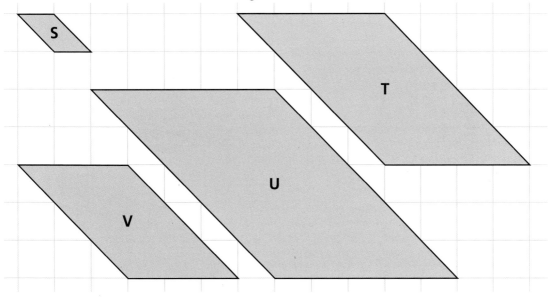

B5 On squared paper draw a copy of shape S above.
Next to it, draw an enlargement of shape S with scale factor 2.

B6 (a) Which one of the shapes P, Q or R is an enlargement of shape M?
(b) What is the scale factor of this enlargement?

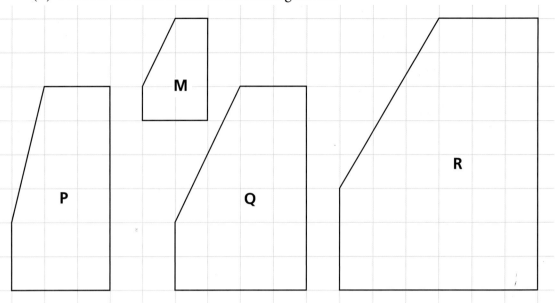

Enlarging designs

Vicky designed this logo for a company.

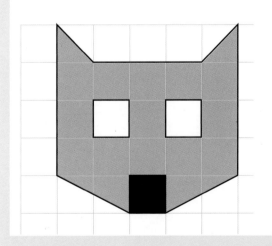

Vicky makes an enlargement of the logo on squared paper.

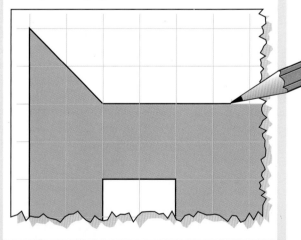

- Draw Vicky's enlarged design on squared paper.
- Design a small logo of your own on squared paper.
  Then make an enlargement of it.
  Choose your own scale factor.

What progress have you made?

Statement

I can recognise enlargements.

Evidence

1 Which one of the shapes A, B or C is an enlargement of shape S?

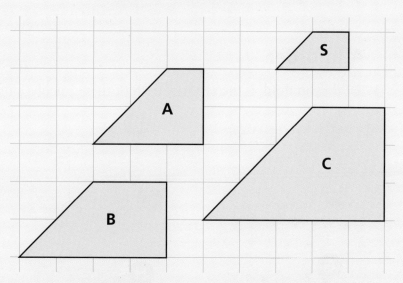

I can work out what scale factor has been used.

2 What is the scale factor of the enlargement in question 1?

I can draw an enlargement using a scale factor that is a whole number.

3 On centimetre squared paper draw

(a) an enlargement of shape A using a scale factor of 2

(b) an enlargement of shape B using a scale factor of 3

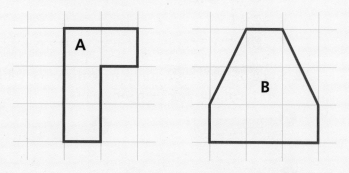

23 Calculate in order

This work will help you
- learn rules to calculate with +, −, × and ÷
- use a calculator effectively

A In order

What do you think is the result of each of these calculations?

A 9 − 5 − 2

B 6 + 4 ÷ 2

C 2 + 2 × 3

D 2 × 5 − 2

E 3 + 1 × 5

F 6 ÷ 2 − 1

G 7 − 6 + 1

H 4 × 6 ÷ 3

I 4 − 4 ÷ 2

J 8 − 3 × 2

K 9 − 3 ÷ 3

Do not use a calculator for questions in section A.

A1 Which of these calculations have 12 as their result?

P   6 + 2 × 3   Q   2 × 2 × 3
R   2 + 1 × 4   S   24 ÷ 3 − 1
T   20 − 4 × 2   U   6 + 12 ÷ 2

A2 Some of these expressions have the same value.
Sort them into five matching pairs.

A 4 × 2 − 5

B 11 − 3 × 2

C 12 ÷ 3 + 1

D 12 + 6 ÷ 2

E 16 ÷ 4 − 1

F 8 + 10 ÷ 5

G 1 + 3 × 2

H 4 + 2 × 3

I 9 − 6 ÷ 3

J 2 × 4 + 7

A3 Work these out.

(a) $6 + 3 - 5$
(b) $2 \times 5 - 3$
(c) $4 + 3 \times 2$
(d) $10 - 2 \times 4$
(e) $10 - 4 - 2$
(f) $12 \div 2 + 1$
(g) $4 + 8 \div 2$
(h) $2 \times 2 \times 3$
(i) $3 + 1 \times 4$
(j) $6 \div 2 + 5$
(k) $12 - 9 \div 3$
(l) $10 + 4 \div 2$
(m) $3 \times 5 - 4$
(n) $12 - 2 \times 5$
(o) $4 - 2 \times 2$

*__A4__ In the expression below, you can replace each diamond with any one of $+$, $-$, $\times$ or $\div$.

8 ◊ 6 ◊ 2

For example, replace the diamonds with $-$ and $\div$ to make $8 - 6 \div 2 = 5$

You must not change the order of the numbers.

Replace each diamond with one of $+$, $-$, $\times$ or $\div$ to make an expression with a value of

(a) 12
(b) 46
(c) 20
(d) 11

Operation 3062

This is a game for two or more players.

Each player needs a game board (sheet 178) and a set of sixteen $+, -, \times, \div$ cards (sheet 180).

- Shuffle your cards and put them face down in a pile in front of you.
- Each player turns over their top two cards. Place your two cards between the 30, 6, and 2 to make the highest result you can.
- Each player says their result. The player with the highest correct result wins two points. (In a draw each player wins two points.) Any other player whose result is correct wins one point.
- Carry on until all the cards have been used.
- The winner is the player with the most points when all the cards have been used.

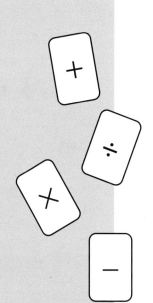

B Brackets

- Always work out expressions in brackets first.

  For example, $2 \times (3 + 4)$
  $= 2 \times 7$
  $= 14$

- If there are no brackets, use these rules to calculate with +, −, × and ÷.

  For example, $3 + 4 \times 2$
  $= 3 + 8$
  $= 11$

  > To calculate with +, −, × and ÷
  > - Multiply or divide **before** you add or subtract.
  > - Otherwise, work from left to right.

B1 Without using a calculator, work these out.

(a) $2 \times (1 + 3)$ (b) $(8 - 2) \div 2$
(c) $(8 - 5) \times 4$ (d) $3 \times (7 - 2)$
(e) $(2 + 8) \div 5$ (f) $2 \times (6 - 5)$
(g) $7 - (4 + 1)$ (h) $(10 + 8) \div 3$
(i) $6 \times (1 + 4)$ (j) $(20 - 6) \div 2$
(k) $8 - (5 - 1)$ (l) $9 \times (1 + 3)$
(m) $12 \div (1 + 3)$ (n) $8 \div (4 - 2)$
(o) $8 \div (12 \div 3)$ (p) $20 \div (10 - 6)$

B2 Some of these expressions have the same value.
Sort them into five matching pairs.

A $(1 + 2) \times 3$ **B** $(24 - 4) \div 5$ **C** $1 + 5 \times 2$

D $1 + 9 \div 3$ **E** $(11 + 1) \div 6$

F $2 \times 9 - 7$ **G** $2 \times (8 - 5)$ **H** $16 - 2 \times 5$

I $7 - 15 \div 3$ **J** $2 + 14 \div 2$

B3 Without using a calculator, work these out.
- (a) $3 \times (5 - 1)$
- (b) $3 \times 5 - 1$
- (c) $(6 + 4) \times 2$
- (d) $6 + 4 \times 2$
- (e) $(10 - 3) \times 3$
- (f) $10 - 3 \times 3$
- (g) $12 - (6 + 1)$
- (h) $12 - 6 + 1$
- (i) $(9 + 6) \div 3$
- (j) $9 + 6 \div 3$
- (k) $(10 - 5) \times 2$
- (l) $10 - 5 \times 2$
- (m) $20 \div (5 - 1)$
- (n) $20 \div 5 - 1$
- (o) $15 \div (3 + 2)$
- (p) $15 \div 3 + 2$
- (q) $16 \div (8 \div 2)$
- (r) $16 \div 8 \div 2$

B4 Each calculation below has a missing number 3, 4 or 6. What is the missing number in each calculation?

- (a) $2 \times \blacksquare \times 2 = 12$
- (b) $\blacksquare \times 3 - 1 = 8$
- (c) $3 \times \blacksquare - 9 = 3$
- (d) $12 \div (\blacksquare - 2) = 3$
- (e) $\blacksquare \div 3 + 5 = 7$
- (f) $10 - \blacksquare \times 2 = 4$
- (g) $1 + 2 \times \blacksquare = 9$
- (h) $4 + 8 \div \blacksquare = 6$
- (i) $\blacksquare - 12 \div 4 = 3$
- (j) $(1 + 2) \times \blacksquare = 9$
- (k) $3 + 2 \times \blacksquare = 15$
- (l) $(10 - \blacksquare) \div 2 = 3$

*__B5__ Find the missing number in each of these calculations.
- (a) $(\blacksquare + 5) \times 2 = 16$
- (b) $\blacksquare + 5 \times 2 = 16$
- (c) $16 - \blacksquare \times 2 = 10$
- (d) $(16 - \blacksquare) \times 2 = 10$
- (e) $(\blacksquare + 12) \div 2 = 12$
- (f) $\blacksquare + 12 \div 2 = 12$
- (g) $6 - \blacksquare \div 3 = 1$
- (h) $(6 - \blacksquare) \div 3 = 1$

C All keyed up

Sue and Jen used a calculator to help with some calculations.

(6 + 13) × 7
= 19 × 7
= 133

Jen used a calculator to work out 19 × 7.

2 + 3 × 17
= 2 + 51
= 53

Sue used a calculator to work out 3 × 17.

C1 Use a calculator to find the value of these.
- (a) (5 + 12) × 6
- (b) 9 × (48 − 19)
- (c) 120 − (63 + 27)
- (d) (102 + 51) ÷ 17
- (e) 241 − (53 − 14)
- (f) 200 ÷ (5 + 20)
- (g) (156 − 12) ÷ 24
- (h) 21 × (20 − 4)
- (i) 143 ÷ (32 − 19)

C2 Use a calculator to find the value of these.
- (a) 14 × 17 − 56
- (b) 36 + 5 × 71
- (c) 67 + 104 ÷ 8
- (d) 16 + 5 × 14
- (e) 126 − 204 ÷ 17
- (f) 11 × 53 − 29
- (g) 158 − 13 × 12
- (h) 131 + 98 ÷ 14
- (i) 126 ÷ 9 + 28

C3 Use a calculator to find the value of these.
- (a) 12 × (71 − 19)
- (b) 63 + 135 ÷ 27
- (c) (912 − 504) ÷ 8
- (d) 231 − 6 × 37
- (e) 126 ÷ (50 − 29)
- (f) 39 × 13 − 205
- (g) 158 − (345 − 217)
- (h) 45 + 225 ÷ 15
- (i) 56 × (7 + 27)

C4 Use a calculator to find the value of these.
- (a) 5 × (3.2 + 1.6)
- (b) 9.7 + 65 × 3
- (c) (241 − 49) ÷ 1.2
- (d) (46 − 28) ÷ 0.5
- (e) 7.8 − 12.1 ÷ 11
- (f) 5.6 × 8 − 2.8
- (g) 1.2 × (3.4 + 4.6)
- (h) 45 + 221 ÷ 34
- (i) (9.1 + 12.6) ÷ 31
- (j) 32 ÷ 64 × 43
- (k) 20 − 1.5 × 7
- (l) 1200 ÷ (2.4 × 20)
- (m) (134 − 93) × 1.8
- (n) 60.9 ÷ (50 − 21)
- (o) 35 × 3.9 − 2.4

What progress have you made?

Statement

I can find the value of an expression with or without brackets.

Evidence

1 Some of these expressions have the same value.
   Sort them into four matching pairs.

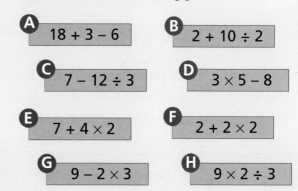

A $18 + 3 - 6$
B $2 + 10 \div 2$
C $7 - 12 \div 3$
D $3 \times 5 - 8$
E $7 + 4 \times 2$
F $2 + 2 \times 2$
G $9 - 2 \times 3$
H $9 \times 2 \div 3$

2 Without using a calculator, work these out.
   (a) $20 - 5 - 1$ (b) $6 + 2 \times 5$
   (c) $15 \div 3 + 1$ (d) $4 + 10 \div 2$
   (e) $9 - 4 \times 2$ (f) $6 - 20 \div 4$

3 Without using a calculator, work these out.
   (a) $4 \times (3 + 1)$ (b) $(6 + 4) \div 5$
   (c) $6 - (3 + 2)$ (d) $(2 + 7) \times 2$
   (e) $(14 - 2) \div 3$ (f) $16 \div (5 + 3)$

4 Without using a calculator, work these out.
   (a) $(3 + 7) \times 2$ (b) $3 + 7 \times 2$
   (c) $(8 - 4) \div 2$ (d) $8 - 4 \div 2$
   (e) $20 \div (4 + 6)$ (f) $20 \div 4 + 6$

I can use a calculator to find the value of an expression.

5 Use your calculator to find the value of these.
   (a) $2 \times (4.1 + 3.4)$ (b) $189 - 12 \times 13$
   (c) $(215 + 65) \div 56$ (d) $136 \div (20 - 3)$
   (e) $13 \times 21 - 46$ (f) $45 + 18 \times 4$
   (g) $3 \times (13.4 - 5.9)$ (h) $89 + 266 \div 19$
   (i) $100 - 472 \div 8$ (j) $500 \div (360 \div 90)$

24 Graphs and charts

This is about displaying information.
The work will help you read and draw graphs and charts.

A Children's income

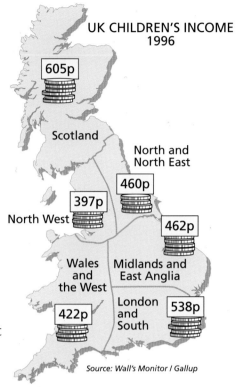

UK CHILDREN'S INCOME 1996

Scotland 605p
North and North East 460p
North West 397p
Midlands and East Anglia 462p
Wales and the West 422p
London and South 538p

Source: Wall's Monitor / Gallup

This picture shows the regions of Great Britain and the results of a survey of children's income.

What information is given for each region?

How does your region compare with other regions?

A1 (a) What is the largest amount of income shown?
(b) Where did children get most income?
(c) Where did children get least income?
(d) What is the difference between the greatest and smallest amounts?

The bar chart shows the results of a survey of boys' and girls' spending.
For example, it shows that about 65% of boys spent money on sweets.

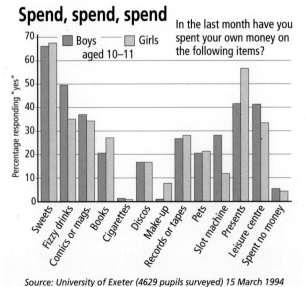

Spend, spend, spend
Boys / Girls aged 10–11
In the last month have you spent your own money on the following items?

Source: University of Exeter (4629 pupils surveyed) 15 March 1994

A2 (a) Roughly what percentage of boys spent money on fizzy drinks?
(b) Roughly what percentage of girls spent money on books?

A3 (a) What was the most popular item that boys spent money on?
(b) What was the most popular item that girls spent money on?
(c) What was the least popular item for girls?

A4 How many pupils were surveyed?

A5 What have you spent your own money on during the last month?

B Shut up!

This graph is about the number of complaints made about noise.

B1 (a) In what year was the highest number of complaints made?

(b) (i) In which two years were there the same number of complaints?

(ii) About how many complaints were made in each of these years?

(c) Between which two years did the number of complaints actually drop?

(d) Between which two years did the number of complaints rise the most?

B2 Complaints went up between 1986 and 1987. By how many roughly?

B3 Copy and complete this story

> Between 1983 and 1985 complaints rose steadily. Then ...

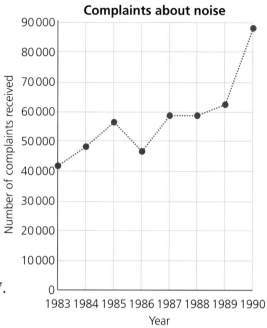

Complaints about noise are investigated. Often the complaints are found to be untrue.

This graph shows the number of complaints found to be true.

B4 Look closely at the numbering on the vertical scale.

Why is there a jagged line at the bottom of the scale, between 0 and 12 000?

B5 Between which two years was there the biggest drop in complaints found to be true?

B6 In what year was there the lowest number of complaints found to be true?

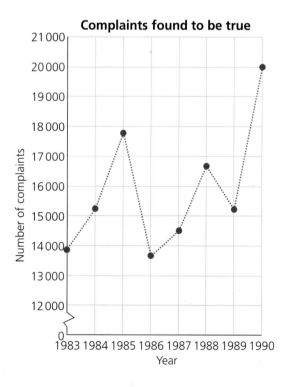

111

C Off the record

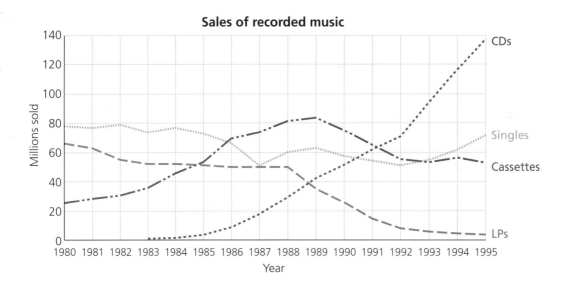

C1 Between 1980 and 1995, the sales of LPs dropped, stayed steady then went down again.

(a) What happened to the sales of CDs?

(b) What happened to the sales of cassettes?

(c) When did CD sales overtake LP sales?

(d) When did cassette sales first reach 70 million?

(e) Roughly how many singles were sold in 1982?

C2 Make up two questions of your own about this graph. Give them to someone else to answer.

Do their answers agree with yours?

D Equal shares

A survey is carried out every few years on who does what at home. Here are the results from the 1991 survey.

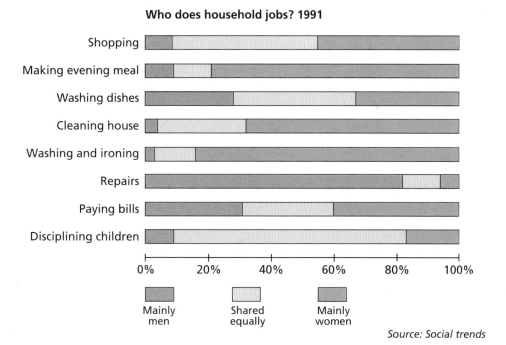

D1 Which job had the highest percentage of
   (a) 'mainly men'
   (b) 'mainly women'

D2 What job was done least by men?
    Explain how you can tell this from the graph.

D3 What jobs were done by 'mainly women' in more than half of all homes?

D4 Which job had the highest percentage of 'shared equally'?

D5 Which jobs were roughly evenly split between men and women?
    Explain how you can tell this from the graph.

E Drawing graphs and charts

Frequency bar charts

Pupils' test scores

37	41	28	60	56
39	17	39	73	64
58	25	44	66	34
32	78	35	46	76
18	39	56	38	75
53	49	55	38	47
53	86	34	64	26
36	22	18	73	9

Drawing frequency bar charts

1. Draw up a tally chart.
2. Fill in tally marks for the data.
3. Fill in the frequency column.
4. Draw and label axes.
5. Draw bars.
6. Write title on chart.

Represent each of these sets of data in a frequency bar chart.
For each one, a suitable tally chart is started.

E1 Ages of people living in a village

41	25	75	64	14	9	23	64
69	33	53	51	27	38	70	22
66	40	38	41	77	80	38	44
13	53	22	59	52	71	60	25

Age group	Tally	Frequency
0–9		
10–19		
20–29		
30–		

E2 Number of cars using a car park each day

54	67	52	74	109	75	86	65
83	100	111	104	67	84	92	74
82	84	89	117	52	95	59	97
92	66	88	94	107	82	57	56
85	98	117	93	68	94	84	77

Number of cars	Tally	Frequency
50–59		
60–69		
70–79		
80–		

E3 Reaction times of pupils using a reaction ruler

20	18	18	13	15	16	17	17
18	22	17	15	11	12	12	13
17	18	21	19	15	16	10	11
16	15	20	16	26	14		

Reaction time (hundredths of a second)	Tally	Frequency
10–12		
13–15		
16–18		
19–		

Line graphs

Temperature of oven in °C at 1 minute intervals from switching on

Time in minutes	0	1	2	3	4	5	6	7	8
Temperature in °C	20	68	129	157	173	184	192	196	198

Drawing line graphs
1. *Draw and label axes.*
2. *Plot points.*
3. *Join up the points.*
4. *Write title of graph.*

Represent these sets of data as line graphs.

E4 This table shows the number of new cases of flu in a fortnight.
Draw your across axis from 0 to 14, and your vertical axis from 0 to 26.

Day	1	2	3	4	5	6	7	8	9	10	11	12	13	14
Cases	3	7	8	11	16	24	22	26	19	15	15	7	3	1

E5 These are the noon temperatures at Rownhams in the first two weeks of December.
Draw your across axis from 0 to 14, and your vertical axis from 0 to 16.

Day in December	1	2	3	4	5	6	7	8	9	10	11	12	13	14
Temperature in °C	15	16	16	12	11	6	3	4	6	8	7	10	9	9

What progress have you made?

Statement

I can read a bar chart.

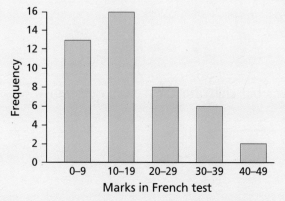

Evidence

1 (a) How many children in this class got less than 10 marks in their French test?
 (b) How many got between 20 and 29 marks?

Statement	Evidence
I can read a line graph.	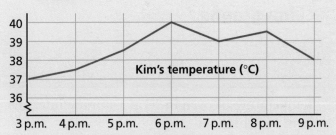

2 (a) What was Kim's temperature at 4:30 p.m?

(b) Describe what happened to Kim's temperature between 3 p.m. and 9 p.m.

I can draw a line graph.

3 This table shows the number of babies born in a hospital in the first 10 days of June.

Day	1	2	3	4	5	6	7	8	9	10
Babies	6	8	7	12	3	4	9	11	12	9

Draw a line graph showing this data.

Make your across axis go from 0 to 10 and your vertical axis go from 0 to 12.

I can use tallying to make a grouped frequency table.

4 Here are the ages of the people in a village.

```
30  23  43  75  63  71  18  25
11   9  66  28   7  22  68  29
55  29  12  40  37  69  70  32
71   8  14
```

Copy and complete this tally chart.

Age group	Tally	Frequency
0–9		
10–19		
20–29		
30–		

I can draw a bar chart.

5 Draw a bar chart for the data in question 4.

25 Fractions 2

This work will help you work out fractions of numbers.

A Chocolate

B Simple fractions of numbers

Halving a number is the same as dividing it by 2

$\frac{1}{2}$ of 6 = 3
6 ÷ 2 = 3

$\frac{1}{2}$ of 8 = 4
8 ÷ 2 = 4

Finding $\frac{1}{4}$ of a number is the same as dividing it by 4

In this chocolate bar, 12 pieces are split into 4 equal groups.

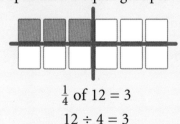

$\frac{1}{4}$ of 12 = 3
12 ÷ 4 = 3

This bar is also split into 4 equal groups.

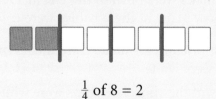

$\frac{1}{4}$ of 8 = 2
8 ÷ 4 = 2

B1 Work these out.

(a) $\frac{1}{2}$ of 4   (b) $\frac{1}{2}$ of 14   (c) $\frac{1}{2}$ of 22   (d) $\frac{1}{2}$ of 10   (e) $\frac{1}{2}$ of 36

B2 Work these out.

(a) $\frac{1}{4}$ of 8   (b) $\frac{1}{4}$ of 32   (c) $\frac{1}{4}$ of 24   (d) $\frac{1}{4}$ of 40   (e) $\frac{1}{4}$ of 52

B3 This chocolate bar shows $\frac{1}{3}$ of 6.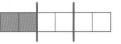
$\frac{1}{3}$ of 6 pieces is 2 pieces.
$\frac{1}{3}$ of 6 = 2

Match each statement below with its chocolate bar.

P $\frac{1}{3}$ of 12 = 4
Q $\frac{1}{6}$ of 18 = 3
R $\frac{1}{2}$ of 12 = 6
S $\frac{1}{4}$ of 12 = 3
T $\frac{1}{3}$ of 18 = 6

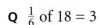
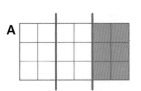
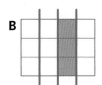
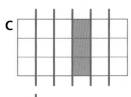
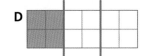
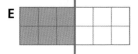

B4 Draw chocolate bars to show each of these.
(a) $\frac{1}{4}$ of 16 = 4 (b) $\frac{1}{3}$ of 24 = 8 (c) $\frac{1}{5}$ of 15 = 3 (d) $\frac{1}{6}$ of 24 = 4

B5 Work these out.
(a) $\frac{1}{2}$ of 16 (b) $\frac{1}{3}$ of 21 (c) $\frac{1}{4}$ of 36 (d) $\frac{1}{5}$ of 10 (e) $\frac{1}{8}$ of 72

B6 Work these out.
Use a calculator if you need to.
(a) $\frac{1}{2}$ of 1564
(b) $\frac{1}{4}$ of 1972
(c) $\frac{1}{3}$ of 23 607
(d) $\frac{1}{7}$ of 366 387
(e) $\frac{1}{8}$ of 2072
(f) $\frac{1}{10}$ of 258 970
(g) $\frac{1}{100}$ of 12 967 000
(h) $\frac{1}{12}$ of 5472
(i) $\frac{1}{6}$ of 3798

B7 About $\frac{1}{100}$ of all females are colour blind.
349 600 females were born in the UK in 1991.
Roughly how many of these babies were colour blind?

B8 The largest biscuit ever made contained 4 million chocolate chips.
$\frac{1}{8}$ of them were white.
How many of the chocolate chips were white?

B9 The longest paper chain ever made had 400 000 links.
One person made $\frac{1}{50}$ of these.
How many links did that person make?

B10 There were 150 000 people at the largest feast ever held.
$\frac{1}{12}$ of these people were vegetarians.
How many vegetarians were at the feast?

C Other fractions of numbers

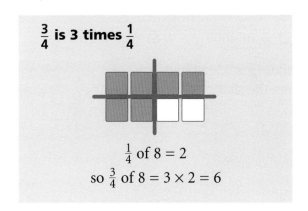

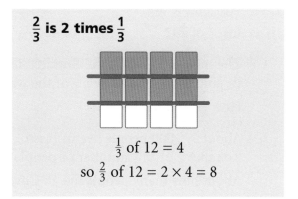

Try to do each of these questions without using a calculator.

C1 Work these out.

(a) $\frac{3}{4}$ of 24   (b) $\frac{3}{4}$ of 32   (c) $\frac{3}{4}$ of 40   (d) $\frac{3}{4}$ of 4   (e) $\frac{3}{4}$ of 44

C2 Work these out.

(a) $\frac{2}{3}$ of 24   (b) $\frac{2}{3}$ of 30   (c) $\frac{2}{3}$ of 18   (d) $\frac{2}{3}$ of 21   (e) $\frac{2}{3}$ of 33

C3 Match each statement below with one of the diagrams.

P $\frac{3}{4}$ of 24 = 18

Q $\frac{3}{5}$ of 20 = 12

R $\frac{2}{3}$ of 18 = 12

S $\frac{3}{4}$ of 20 = 15

T $\frac{5}{6}$ of 24 = 20

U $\frac{5}{6}$ of 18 = 15

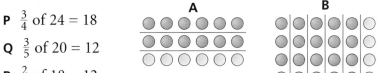
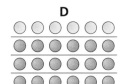
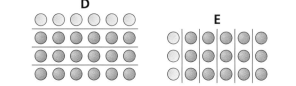

C4 Work these out.

(a) $\frac{3}{4}$ of 36   (b) $\frac{2}{5}$ of 30   (c) $\frac{3}{8}$ of 40   (d) $\frac{5}{6}$ of 42   (e) $\frac{2}{3}$ of 60

C5 Leontina Albina, who lived in Chile, had 40 surviving children. $\frac{2}{5}$ of them were girls. How many were girls?

C6 Jack is 180 cm tall and 36 years old.

(a) Zoe is $\frac{8}{9}$ of Jack's height. How tall is Zoe?

(b) Zoe's age is $\frac{2}{3}$ of Jack's age. How old is Zoe?

Use a calculator for these questions if you need to.

C7 Work these out.
(a) $\frac{2}{3}$ of 792
(b) $\frac{4}{5}$ of 4610
(c) $\frac{7}{10}$ of 4000
(d) $\frac{3}{4}$ of 492 000

C8 The main lift shaft at the Empire State Building is 320 m high.
In January 2000 a lift fell $\frac{2}{5}$ of the way down the shaft. No one was hurt!
How far did the lift fall?

C9 About 375 000 people live in Bristol.
Roughly $\frac{4}{5}$ of this number of people live in Coventry.
About how many people live in Coventry?

C10 The area of France is 544 000 km².

(a) The area of Ireland is $\frac{1}{8}$ the area of France.
What is the area of Ireland?

(b) The area of Finland is $\frac{5}{8}$ the area of France.
What is the area of Finland?

C11 A small beech tree has approximately 2000 leaves.
$\frac{3}{10}$ of the leaves fell off one autumn. About how many leaves fell?

What progress have you made?

Statement

I can work out simple fractions of numbers.

Evidence

1 Work these out.
(a) $\frac{1}{2}$ of 18
(b) $\frac{1}{4}$ of 20
(c) $\frac{1}{5}$ of 30

2 Work these out.
(a) $\frac{1}{8}$ of 72
(b) $\frac{1}{7}$ of 35
(c) $\frac{1}{6}$ of 60

3 Use a calculator to work these out.
(a) $\frac{1}{4}$ of 220
(b) $\frac{1}{5}$ of 1000
(c) $\frac{1}{6}$ of 756

I can work out more difficult fractions of numbers.

4 Work these out.
(a) $\frac{2}{3}$ of 21
(b) $\frac{3}{4}$ of 80
(c) $\frac{3}{5}$ of 35

5 Use a calculator to work these out.
(a) $\frac{3}{7}$ of 300 272
(b) $\frac{4}{11}$ of 7392
(c) $\frac{3}{5}$ of 855
(d) $\frac{7}{12}$ of 1440

Negative numbers

This work will help you
- put negative numbers in order
- find differences between negative numbers
- use graphs and charts with negative numbers

A Lift-off!

The most important event for a rocket is lift-off. Times are measured from lift-off.

T minus 30 minutes

T plus 30 minutes

This is 30 minutes **before** lift-off. We write this –30 min.

This is 30 minutes **after** lift-off. We write this +30 min.

Here are some events involving the Space Shuttle.

–3 hours Crew leave for shuttle

–9 minutes Decision to 'GO' or not

–2 hours 30 minutes Crew strapped into the cabin

–1 hour 20 minutes Cabin hatch closed

+2 minutes Shuttle 40 km high, booster rockets are used up and fall to Earth

+7 seconds Shuttle clears the launch tower

0 Lift-off

–1 hour 10 minutes Cabin leak check

In which order do these events happen?

Timeline marks: +1 hour, 0, –1 hour, –2 hours, –3 hours

Seconds away!

Here are some more Shuttle events.
Can you put these in order?

Time	Event	Letter
+26 seconds	Reduce engines' power	R
−50 seconds	Ground power removed	G
−3 seconds	Put engines on 90% power	P
−10 seconds	Computer begins engine start sequence	C
−31 seconds	On-board computer takes control	O
+30 seconds	Full power	F

B Freezing cold

Temperature is measured in degrees Celsius (°C)

In cold parts of the world temperatures fall below 0°C.

Water freezes at 0°C.

Temperatures below 0°C are below freezing for water!

Temperature records

Coldest recorded in the UK:
10 January 1982, Braemar, Scotland   −27°C

Hottest recorded in the UK:
3 August 1990, Cheltenham, England   37°C

Coldest recorded in the world:
21 July 1983, Vostok, Antarctica   −89°C

Hottest recorded in the world:
September 1922, Libya   58°C

B1 (a) Write down the temperature in each of these cities.

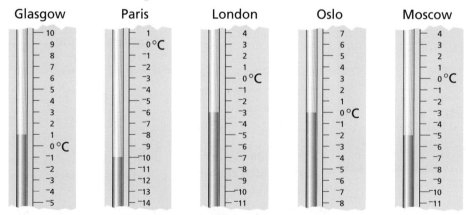

(b) Which city has the coldest temperature?

(c) Which city has the warmest temperature?

Use the thermometer on the right to help you answer these questions.

B2 Which is the colder temperature, ⁻2°C or ⁻4°C?

B3 Which of these temperatures is lowest? ⁻12°C, ⁻4°C, 2°C, ⁻8°C

B4 What temperature is 4 degrees lower than 7°C?

B5 What temperature is 4 degrees lower than 3°C?

B6 What temperature is 5 degrees lower than 1°C?

B7 What temperature is 4 degrees lower than 0°C?

B8 What temperature is 5 degrees lower than ⁻3°C?

B9 What temperature is 5 degrees higher than ⁻3°C?

B10 Here are some thermometers.

(a) Write down the temperature shown on each one.

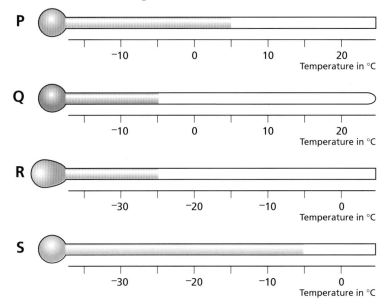

Food which is kept cool will stay fresh longer.
- A normal fridge should keep food between 5°C and 0°C.
- Freezers should keep food at or below ⁻18°C.

(b) Which of the thermometers P, Q, R and S show a temperature suitable for a freezer?

B11 Say whether these are a suitable temperature for a freezer, fridge or neither.

(a) ⁻25°C   (b) 3°C   (c) ⁻15°C   (d) 1°C

B12 This map shows the noon temperature at some cities in Europe on a January day.

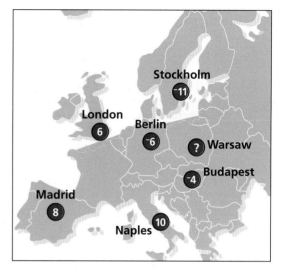

(a) Which city was warmer, Berlin or Budapest?

(b) How many degrees warmer was Madrid than Berlin?

(c) How many degrees warmer was London than Stockholm?

(d) Warsaw was 3 degrees warmer than Stockholm.

What was the temperature in Warsaw?

B13 Make this sentence true by putting the numbers below into the correct spaces.

A temperature of .... °C is .... degrees higher than a temperature of .... °C.

B14 Make this sentence true by putting the numbers above into the spaces.

A temperature of .... °C is .... degrees lower than a temperature of .... °C.

B15 These are the temperatures at midnight in Helsinki for the first fortnight in May.

May	1	2	3	4	5	6	7	8	9	10	11	12	13	14
Temp. (°C)	−10	−8	−7	−8	−3	1	1	2	3	2	3	−1	−1	2

(a) On how many days was the midnight temperature above freezing?

(b) How many degrees difference is there between the midnight temperature on 1st May and 9th May?

B16 This table shows the temperatures at noon and midnight for a town in Canada for the first fortnight in April.

April	1	2	3	4	5	6	7	8	9	10	11	12	13	14
Noon temp. (°C)	−10	−12	−12	−6	0	0	2	3	5	−1	−2	−1	3	5
Midnight temp. (°C)	−21	−18	−19	−14	−9	−8	−5	−2	2	−5	−7	−5	−2	1

Which of these statements are true?

(a) For half the fortnight noon temperatures were below freezing.

(b) It was always freezing at midnight.

(c) If it was freezing at noon, it was always freezing at midnight.

(d) On April 2nd it was 6 degrees colder at midnight than at noon.

C Temperature graphs

C1 This bar chart shows mean monthly temperatures at noon in Wellington.

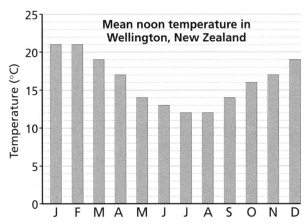

(a) In which months was the mean temperature at noon hotter than 20°C?

(b) In which months was the mean temperature at noon colder than 14°C?

(c) What is the difference in mean noon temperature between January and August?

C2 This line graph shows the mean noon and midnight temperatures in New York.

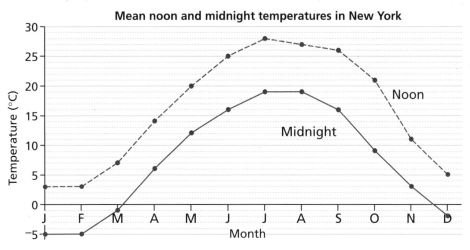

(a) What is the mean noon temperature in June?

(b) What is the mean midnight temperature in January?

(c) Which months have the lowest mean midnight temperature?

(d) What is the difference between the mean noon and midnight temperatures in March?

(e) Look at the mean temperatures in December.
How many degrees higher is the mean at noon than at midnight?

(f) Which month has the biggest difference between the mean temperatures at noon and midnight?

What progress have you made?

Statement

I can put negative numbers in order.

I can find differences between negative numbers.

I can read graphs which include negative numbers.

Evidence

1. Put these temperatures in order, lowest first.

    2°C ⁻3°C ⁻32°C 15°C ⁻14°C ⁻1°C

2. These are the noon temperatures on a January day at some cities in the USA.

    Boston   ⁻12°C   Chicago   ⁻7°C
    Miami   21°C   Los Angeles 16°C

    (a) How many degrees warmer was Los Angeles than Chicago?

    (b) How many degrees warmer was Chicago than Boston?

    (c) Denver was 20°C warmer than Boston. What was the temperature in Denver?

3. This graph shows the temperature in °C on a winter's night somewhere in England.

    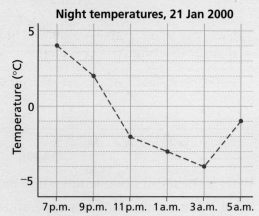

    (a) What temperature was it at 1 a.m.?

    (b) How many degrees difference was there between the temperatures at 7 p.m. and 11 p.m.?

    (c) Between which two times did the temperature change the most?

    (d) At 7 a.m. the temperature had risen by 3 degrees from 5 a.m.
    What was the temperature at 7 a.m.?

Review 4

1 Each photo must be in the middle of the card.
  What size should the borders be?

  (a) 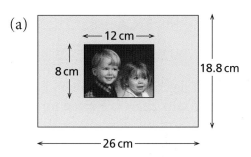   (b)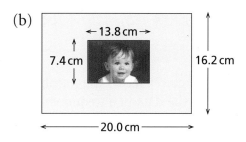

2 What fraction of each tile is red?
  What fraction is yellow?

  (a) 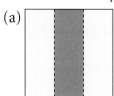   (b) 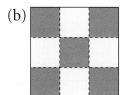   (c) 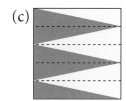   (d)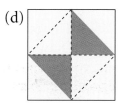

3 On squared paper, draw an enlargement of each of these shapes.
  The scale factor is different for each one.

  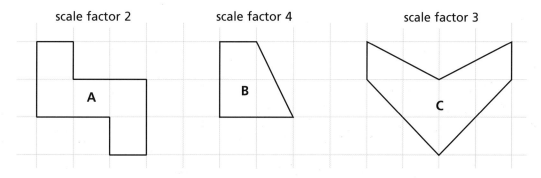

4 Work out these in your head.
  (a) $\frac{1}{2}$ of 12   (b) $\frac{1}{2}$ of 30   (c) $\frac{1}{3}$ of 24   (d) $\frac{1}{4}$ of 24   (e) $\frac{1}{5}$ of 35
  (f) $\frac{1}{3}$ of 15   (g) $\frac{2}{3}$ of 15   (h) $\frac{1}{5}$ of 30   (i) $\frac{2}{5}$ of 30   (j) $\frac{3}{5}$ of 30

5 Harry is 12 years old and 150 cm tall. Harriet is half as old and $\frac{2}{3}$ as tall.
  How old and how tall is Harriet?

6 Without using a calculator, work these out.

(a) $4 \times (5 - 2)$ (b) $4 \times 5 - 2$ (c) $24 \div 4 - 2$ (d) $24 \div (4 - 2)$

(e) $32 + 8 \div 4$ (f) $(32 + 8) \div 4$ (g) $20 - 6 - 2$ (h) $20 - (6 - 2)$

7 Find the missing number in each of these.

(a) $10 + (\blacklozenge - 3) = 15$ (b) $2 \times (\blacktriangledown + 1) = 12$ (c) $3 \times \blacktriangle + 6 = 12$

(d) $6 + \blacksquare \times 2 = 14$ (e) $18 - \square \div 2 = 8$ (f) $\blacktriangledown + 5 \times 2 = 20$

8 The data set shows the number of birds seen each day on a bird table in February. Show the data in a frequency bar chart. A suitable tally chart is started for you.

40	15	35	44	4	9	17
50	23	33	7	10	18	55
38	40	29	11	7	20	38
13	23	42	39	22	51	48

Number of birds	Tally	Frequency
0–9		
10–19		
20–		

9 This table shows the dawn temperature in Seville in the first 10 days of March. Draw a line graph to show the data.

Make your across axis go from 0 to 10, and your vertical one from 0 to 20.

Day	1	2	3	4	5	6	7	8	9	10
Temperature (°C)	2	5	0	12	10	18	16	11	5	8

10 What temperature is

(a) 4 degrees higher than 14°C

(b) 4 degrees lower than 14°C

(c) 10 degrees higher than ⁻4°C

(d) 10 degrees lower than ⁻4°C

(e) 5 degrees lower than 0°C

(f) 10 degrees higher than ⁻10°C

11 Read this entry from Guérin's diary and then answer the questions below.

> I got up at a quarter to 7 and it was ⁻5°C. Brrrr! When I had lunch (at half past 12) the temperature was 8°C. Much better. But at tea-time at 5:15 p.m. it was back down to ⁻2°C. Tea took 20 minutes and afterwards I went skating for an hour and a half!

(a) How long was it between when Guérin got up and when he had lunch?

(b) By how much did the temperature rise from when Guérin got up to lunch-time?

(c) How long was it between lunch-time and tea-time?

(d) What was the difference in temperature between lunch-time and tea-time?

(e) At what time did Guérin finish skating?